CW01370536

EURO Advanced Tutorials on Operational Research

Series Editors

M. Grazia Speranza, Brescia, Italy
José Fernando Oliveira, Porto, Portugal

More information about this series at
http://www.springer.com/series/13840

Renata Mansini • Włodzimierz Ogryczak •
M. Grazia Speranza

Linear and Mixed Integer Programming for Portfolio Optimization

Springer

Renata Mansini
Department of Information Engineering
University of Brescia
Brescia, Italy

Włodzimierz Ogryczak
Institute of Control and
 Computation Engineering
Warsaw University of Technology
Warsaw, Poland

M. Grazia Speranza
Department of Economics and Management
University of Brescia
Brescia, Italy

EURO Advanced Tutorials on Operational Research
ISBN 978-3-319-18481-4   ISBN 978-3-319-18482-1   (eBook)
DOI 10.1007/978-3-319-18482-1

Library of Congress Control Number: 2015942915

Springer Cham Heidelberg New York Dordrecht London
© Springer International Publishing Switzerland 2015
This work is subject to copyright. All rights are reserved by the Publisher, whether the whole or part of the material is concerned, specifically the rights of translation, reprinting, reuse of illustrations, recitation, broadcasting, reproduction on microfilms or in any other physical way, and transmission or information storage and retrieval, electronic adaptation, computer software, or by similar or dissimilar methodology now known or hereafter developed.
The use of general descriptive names, registered names, trademarks, service marks, etc. in this publication does not imply, even in the absence of a specific statement, that such names are exempt from the relevant protective laws and regulations and therefore free for general use.
The publisher, the authors and the editors are safe to assume that the advice and information in this book are believed to be true and accurate at the date of publication. Neither the publisher nor the authors or the editors give a warranty, express or implied, with respect to the material contained herein or for any errors or omissions that may have been made.

Printed on acid-free paper

Springer International Publishing AG Switzerland is part of Springer Science+Business Media (www.springer.com)

To Giovanni and Matteo
To Barbara and Bartek
To Massimo, Chiara, and Laura

Preface

Portfolio theory was first developed by Harry Markowitz in the 1950s. His work, which was extended by several researchers, provides the foundation of the so-called *modern portfolio theory*. Markowitz work has been published and discussed in several papers and books. He introduced the concept of diversification and captured in a model the importance of investing in a diversified portfolio. His pioneering model has the goal to find the optimal trade-off between the risk and the return of an investment. The risk of the portfolio is measured through the variance of the portfolio rate of return. The resulting optimization model, which is a single period model for portfolio selection, is quadratic. In the last two decades, several models have been proposed for portfolio optimization that use different (with respect to the variance) functions to measure the performance of a portfolio. Several resulting optimization models are linear. Linear programming models open up the possibility to consider additional features of the investment problem, also those that imply the introduction of binary and integer variables.

The optimization models for portfolio selection that are the focus of this book are of interest in many application areas, from finance to business and engineering. Although we use the basic language of finance and talk about the investment of a capital in assets, the models can be applied to a broad variety of portfolio selection problems.

The book presents the general problem of single period portfolio optimization, the different linear models arising from different performance measures, and the mixed-integer linear models resulting from the introduction of real features. Transaction costs may be charged when a specific asset or a group of assets is included in the portfolio. The presence of such costs is among the most important features of the real problems that, to be embedded in an optimization model, require in most cases binary variables. Other linear models, such as models for portfolio rebalancing and index tracking, are also covered. Computational issues are discussed and the theoretical framework, including the concepts of risk averse preferences, stochastic dominance, and coherent risk measures, is provided.

The material in the book is presented assuming that the reader does not have a background in finance or in portfolio optimization. Only some experience in linear and mixed-integer models is assumed. The material is presented in a didactic way. Concepts are accompanied by comments and examples. This is in a sense the book we wish we had when we first started working on portfolio optimization, coming from education in operations research. The target readers of this book are students – undergraduate, MBA, PhD – researchers, and professionals interested in the use of optimization models for portfolio selection.

Brescia, Italy	Renata Mansini
Warsaw, Poland	Włodzimierz Ogryczak
Brescia, Italy	M. Grazia Speranza

Acknowledgments

The authors acknowledge the precious help of Enrico Angelelli, Claudia Archetti, and Gianfranco Guastaroba who read previous versions of some chapters and suggested various ways to improve them.
Włodzimierz Ogryczak acknowledges the research support of the National Science Centre (Poland) under the Grant DEC-2012/07/B/HS4/03076.

Contents

1 Portfolio Optimization 1
   1.1   Introduction 1
   1.2   Market and Diversification 1
   1.3   The Optimization Framework 3
   1.4   Portfolio Performance 5
   1.5   Basic Concepts and Notation 7
   1.6   Markowitz Model 9
   1.7   Risk and Safety Measures 12
   1.8   Handling Bi-Criteria Optimization Problems 14
   1.9   Notes and References 17

2 Linear Models for Portfolio Optimization 19
   2.1   Introduction 19
   2.2   Scenarios and LP Computability 19
   2.3   Basic LP Computable Risk Measures 21
   2.4   Basic LP Computable Safety Measures 27
   2.5   The Complete Set of Basic Linear Models 32
       2.5.1   Risk Measures from Safety Measures 33
       2.5.2   Safety Measures from Risk Measures 35
       2.5.3   Ratio Measures from Risk Measures 36
   2.6   Advanced LP Computable Measures 38
   2.7   Notes and References 44

3 Portfolio Optimization with Transaction Costs 47
   3.1   Introduction 47
   3.2   The Structure of Transaction Costs 48
   3.3   Accounting for Transaction Costs in Portfolio Optimization 53
   3.4   Optimization with Transaction Costs 59
   3.5   A Complete Model with Transaction Costs 60
   3.6   Notes and References 61

4 Portfolio Optimization with Other Real Features ... 63
- 4.1 Introduction ... 63
- 4.2 Transaction Lots ... 64
- 4.3 Thresholds on Investment ... 67
- 4.4 Cardinality Constraints ... 70
- 4.5 Logical or Decision Dependency Constraints ... 71
- 4.6 Notes and References ... 71

5 Rebalancing and Index Tracking ... 73
- 5.1 Introduction ... 73
- 5.2 Portfolio Rebalancing ... 74
- 5.3 Index Tracking ... 78
  - 5.3.1 Market Index ... 79
  - 5.3.2 An Index Tracking Model ... 80
- 5.4 Enhanced Index Tracking ... 83
- 5.5 Long/Short Portfolios ... 85
- 5.6 Notes and References ... 86

6 Theoretical Framework ... 87
- 6.1 Introduction ... 87
- 6.2 Risk Averse Preferences and Stochastic Dominance ... 88
- 6.3 Stochastic Dominance Consistency ... 91
- 6.4 Coherent Measures ... 93
- 6.5 Notes and References ... 95

7 Computational Issues ... 97
- 7.1 Introduction ... 97
- 7.2 Solving Linear and Mixed Integer Linear Programming Problems ... 99
- 7.3 A General Heuristic: The Kernel Search ... 101
- 7.4 Issues on Data ... 103
- 7.5 Large Scale LP Models ... 108
- 7.6 Testing and Comparison of Models ... 111
- 7.7 Notes and References ... 113

References ... 115

Chapter 1
Portfolio Optimization

1.1 Introduction

The problem of investing money is common to citizens, families and companies. Families often simply aim at protecting their savings from inflation, or at gaining some additional money but without putting their savings at risk. Companies face more complex investment strategies looking for high return opportunities at a reasonable risk. Financial institutions invest money on behalf on any investor. Over the years the number of investment opportunities has increased dramatically, because of the globalization of financial markets and also because of the creation of new investment instruments. Nowadays, major classes of instruments include bonds, equities or stocks, investment funds (including the recent socially responsible funds), derivatives, assets created from securitization processes (for example, asset-backed securities). Most of the concepts dealt with in this book can be applied to the majority of investment instruments. Nevertheless, we mainly refer to equities and investment in the stock exchange market.

1.2 Market and Diversification

Any specific financial instrument that is tradable is called *asset*. We consider the problem of a private or institutional investor interested in investing a certain amount of money, called *capital*, in a specified set of assets, called *available assets*. Building a *portfolio* means deciding the amount of capital, called *share*, to invest in each available asset. A null share indicates that the asset is not selected.

Assets may be evaluated individually, on the basis of their past and expected future performance. Clearly, an asset that performed well in the recent past is attractive. It may be expected that this asset will perform well in the future. However, in this case the capital should be all invested in the asset with the best performance. The problem is that the future is uncertain. There is no guarantee that an asset that performed well in the last month will keep performing well in the next one.

To evaluate an asset performance we look at its *quoted price* or *quotation*, also called *market price*, as determined by the market. This is the economic value at which an asset is bought or sold. The estimation of future quotations of an asset is a hard task. The quoted price is the result of several components and factors such as strength and stability of the underlying company, its profitability (and therefore expected dividends), future projections on the sector in which the company operates. Each investor may assign a different economic value to an asset, depending on the possessed information and expectations. This implies that different investors are willing to sell or buy at different values of an asset quotation. An investor acting rationally will not sell an asset at a price lower than estimated or purchase at a higher price. The presence of different investors facilitates an increase of market liquidity. The wider the range of economic values that investors attribute to an asset, the more numerous will be the individual situations of convenience to sell or buy at market price, and therefore the higher the number of exchanges for that asset. Information on the asset quotations is public and available to anyone.

If many will invest in the same asset the market price of that asset will increase and, as a consequence, the asset will become less attractive, at least for some investors. The future performance of an asset is, thus, the result of the whole set of the trading activities, that depend on past performance of the assets but also on investor expectations on future performance, on the evolution of the different sectors, on the situation of the underlying companies, on economy and management. If banks go through a crisis the assets of the banking sector will perform poorly. If a car manufacturer creates a new successful model the performance of the asset will probably improve. Similarly, if the CEO (Chief Executive Officer) of a company will be replaced by a well known successful CEO, the quotation of the associated asset will increase.

Also different macroeconomic variables that are related to national and international economies or to monetary and fiscal policies may affect the future earnings of a company and, thus, the expectations on its asset quotation. For example, information on the GDP (Gross Domestic Product) of a country may have an impact on the assets of companies based in the country. Besides, quotations are sensitive to interest rates. An increase of interest rates may lead to an increase in costs for a company, and thus to a drop in earnings, lowering expectations on future dividends and, hence, causing a reduction in quotation. This is true for almost all investment sectors, but for insurance and banking assets for which the effect is opposite. Other macroeconomic variables that may affect asset quotations are the inflation rate, that impacts the expected growth of the economy, and the exchange rate, that cannot be ignored if the investor wants to operate in a global market. Also economic and monetary policies adopted by a government and its central bank may have a strong effect on asset performance. Unforeseen events may take place and affect quotations. Some may be predicted on the basis of the current information, some may be hard or even impossible to predict.

1.3 The Optimization Framework

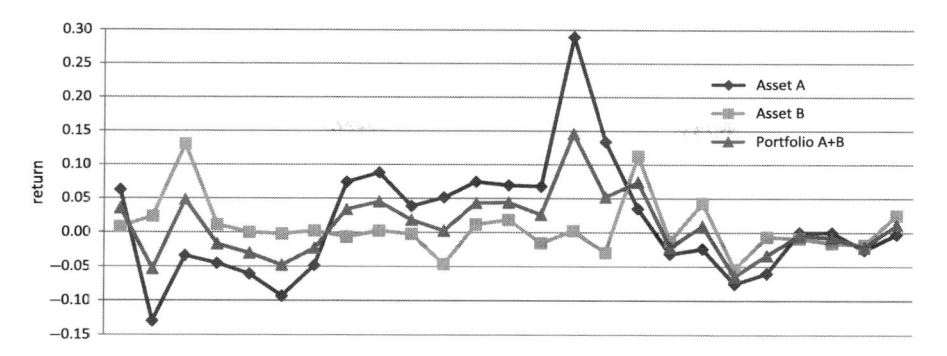

Fig. 1.1 The diversification effect

Investing the whole capital in a single asset is a highly risky action as, in case the asset performance will be different than expected, not only there may be no gain but part of the invested capital may get lost. If the portfolio includes different assets it is unlikely that the performance of all will drop, especially if the assets belong to different sectors (e.g., goods vs services, old vs new economy assets), to different countries or geographic areas. The portfolio should be built in such a way that, even if the performance of all the assets will drop, the minimum possible capital will be lost. Usually, the quotations of assets of the same sector tend to move in the same direction, upward or downward. A portfolio can take advantage of the fact that the performance of assets of different sectors may move in opposite directions, one upward and the other downward, or viceversa. In general, there is a non-null correlation between the performance of pairs of assets. The concept of spreading investment over a set of assets is called *diversification*. Diversification reduces volatility of portfolio performance since not all asset quotations move up or down at the same time or at the same rate. In Fig. 1.1 we show the effect of diversification over two assets A and B. The monthly rate of return series of each asset and of a portfolio equally investing in the two assets are shown over 2 years. Note how diversification allows a more stable performance.

1.3 The Optimization Framework

In order to define the optimization framework, timing is an issue to be carefully considered. There is a moment where the portfolio is built for the first time and assets are only bought. We call it the *investment time*. There may be subsequent moments where the existing portfolio is revised and adjusted to market changes. In this case, assets may be bought or sold. Such adjustments are made at fixed or variable intervals of time and might be desirable although implying additional transaction costs. We focus on the problem of building a portfolio at the investment time and address the so called rebalancing problem in Chap. 5.

```
        ┌─────────────────────┐
   ┌───▶│   identification    │
   │    │  of available assets│
   │    └──────────┬──────────┘
   │               ▼
   │    ┌─────────────────────┐
   │ ┌─▶│     Estimation      │
   │ │  │ of future performance│
   │ │  │  of available assets │
   │ │  └──────────┬──────────┘
   │ │             ▼
   │ │  ┌─────────────────────┐
   │ │┌▶│    Optimization     │
   │ ││ │       model         │
   │ ││ └──────────┬──────────┘
   │ ││            ▼
   │ ││ ┌─────────────────┐    ┌─────────────────┐
   └─┴┴─│  Ex-post analysis│──▶│ Implementation  │
        │                 │    │   of solution   │
        └─────────────────┘    └─────────────────┘
```

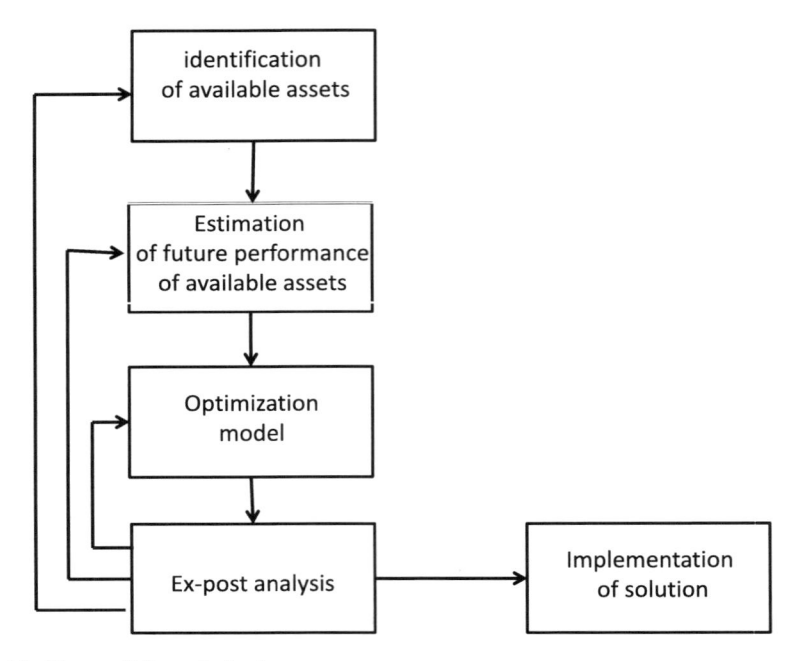

Fig. 1.2 The portfolio optimization process

We assume that the investor has in mind the period of time over which the portfolio will be kept unchanged. This period may be relatively short or quite long, usually of the order of months or years. The investor is interested in the wealth at the end of this time period. The expectations on future performance of the assets will be referred to that moment in time, that is called *target time*. The strategy where a portfolio is built at the investment time and kept until the target time is called *buy and hold* strategy.

The process of building a portfolio through a scientific approach can be described as a sequence of phases (see Fig. 1.2):

1. Identification of the set of available assets;
2. Collection of information, beliefs, methods to estimate the performance of available assets at the target time;
3. Selection of a model that, on the basis of the estimations obtained in the previous phase, generates a portfolio;
4. Assessment of the model through ex-post analysis and possible feedback to one of the previous phases;
5. Implementation of the portfolio.

In this book we mainly focus on phase 3 and present models for *portfolio optimization*. We will address the other phases in Chap. 7.

In the portfolio optimization process there are several aspects and characteristics that may be relevant to the decision of which portfolio to build. While some are very difficult to capture in a model, some may be introduced at the cost of making the model more complex and computationally harder to solve. Transaction costs, charged by a broker or a financial institution to trade assets, are an example of the latter case. We will initially present optimization models for the most basic case and discuss additional portfolio features in Chaps. 3 and 4.

1.4 Portfolio Performance

The concept of performance of an asset and of a portfolio is crucial for the design of an optimization model. What performance measure can we use? It is relatively simple to measure the past performance. An asset or a portfolio performed well over a certain period of time if its return is satisfactory when compared, for example, with benchmarks, such as market indices.

The *rate of return* r_t of an asset in a unit time interval beginning at time $t-1$ and ending at time t is defined as $r_t = (q_t - q_{t-1})/q_{t-1}$, where q_{t-1} and q_t are quotations at times $t-1$ and t, respectively. The rate of return measures the appreciation, if $r_t > 0$, or depreciation, if $r_t < 0$, of the asset quotation during the period $(t-1, t)$. Note that r_t is never lower than -1. In general, the computation of the rate of return may also include possible dividends paid in the period $(t-1, t)$. In the following we will not consider dividends explicitly. In order to obtain at time t the return, that is the amount of money earned or lost over the time period $(t-1, t)$, the investor multiplies the capital invested at time $t-1$ by r_t.

Assessing the future performance of an asset is a more difficult task. The expected rate of return at the target time cannot be the only performance measure for an asset or a portfolio. The problem is all contained in the word 'expected', in the lack of guarantee that the expected return will be achieved. An asset rate of return at the target time can be modelled as a random variable and, as a consequence, also the rate of return of a given portfolio, that is obtained from the rates of return of the assets that compose it, is a random variable. Let us consider a case where assets may take at the target time any of 25 possible values, each with the same probability. In Fig. 1.3 we show the 25 possible rates of return of two portfolios at the target time. The two portfolios have the same expected rate. Which one should be preferred? Risk averse investors would probably choose the portfolio that has the most reliable performance, that is the one with the least variability of the rate of return around the expected value. This would reduce the risk of a large loss, although it would reduce at the same time the probability of a large gain. In Fig. 1.4 the case of two portfolios with different expected rates of return and the same variability pattern is shown. In this latter case, the portfolio with the largest expected rate of return should be clearly preferred.

The performance issue is related to the trade-off between the expected value of the uncertain rate of return of a portfolio and its level of variability. The variability

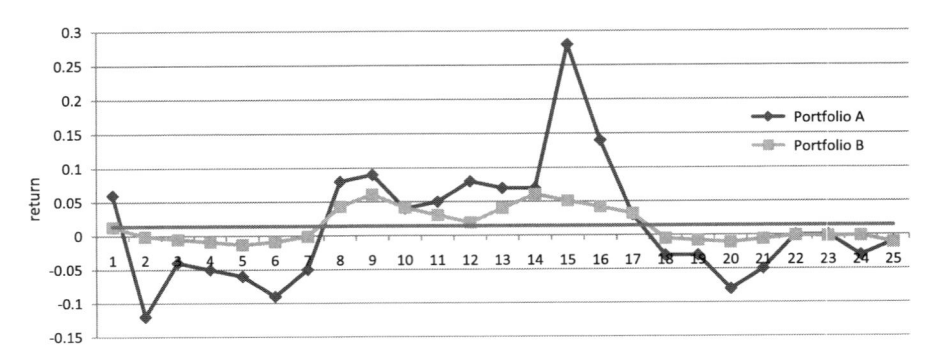

Fig. 1.3 The possible rates of return of two portfolios with identical expected value

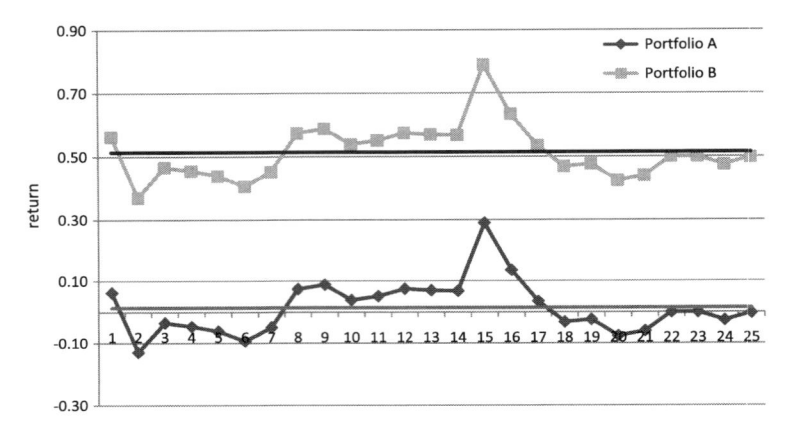

Fig. 1.4 The possible rates of return of two portfolios with identical variability profile

should measure the associated *risk*, the risk of not gaining what is expected. Usually, a high expected return cannot be achieved without a high level of risk. The choice of a specific trade-off level, for example, between a portfolio with high return and high risk and one with low return and low risk, can only be performed by the investor. The risk aversion will make an investor prefer the latter to the former and the risk propensity the former to the latter.

Measuring the risk is a crucial aspect of portfolio optimization and will be treated in depth in this book. Let us assume, for the moment, that we have chosen a measure and can, thus, compute the expected return and risk of any portfolio. The best possible trade-offs between the two objectives can be expressed through the risk/return frontier. In Fig. 1.5 we show an example, where each point of the curve represents a portfolio. Portfolio A has lower risk and lower expected return than portfolio B. On the other hand, portfolios A and B should both be preferred to portfolio C because the latter has higher risk and the same expected return of A and the same risk and lower return than B. Portfolio C is said to be *dominated* by A and B.

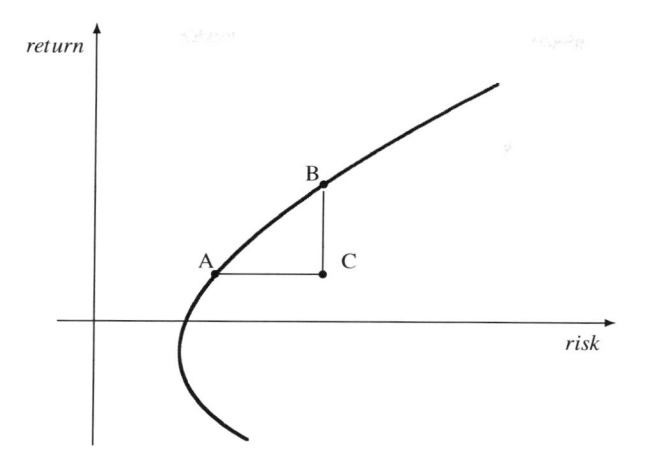

Fig. 1.5 The risk/return frontier

The risk is a natural way to express the uncertainty of the return. However, a different kind of measure can be used instead of the risk. Such a measure is the *safety* and is related to the probability of achieving a certain level of return. If the uncertainty is captured by a risk measure the investor will aim at minimizing the risk. Viceversa, the investor will aim at maximizing a safety measure. Several safety measures will be analyzed in this book. Later in this chapter, in Sect. 1.7, we will discuss relations between risk and safety measures.

1.5 Basic Concepts and Notation

The portfolio optimization problem is based on a single investment period. At the investment time an investor allocates the capital, assigning a share to each available asset. We can express the share of an assets as a percentage of the capital. In this case we do not need to specify the value of the capital in the formulation of the portfolio optimization models. We call such a percentage the *weight* of the asset in the portfolio.

During the investment period, each asset generates a rate of return. Given the weights of the assets, the portfolio rate of return is the weighted average of the rates of return of the assets and may be negative, null or positive. Thus, during the investment period the value of the capital has changed and at the target time may be smaller, equal or higher than the initial invested capital, depending on the sign of the return.

Example 1.1 Let us consider a set of available assets and a capital equal to 1,000 € and suppose that two assets are selected to be part of the portfolio with shares 400 € and 600 €, respectively. The expected rate of return of the two assets at the target time, suppose 1 year after, is 3 % and 5 %, respectively. Then, the expected value of the capital at the target time is $(1 + 0.03) * 400 + (1 + 0.05) * 600 = 1{,}042$. The expected rate of return of the portfolio is, thus, $(1{,}042 - 1{,}000)/1{,}000 = 4.2\,\%$.

The weights of the assets are $400/1{,}000 = 0.4$ and $600/1{,}000 = 0.6$. The expected rate of return of the portfolio is, thus, $0.03 * 0.4 + 0.05 * 0.6 = 4.2\,\%$. When the value of the capital C is given, the expected value of the capital at the target time is $(1 + 0.042) * C$. With $C = 1{,}000$ the value is $1{,}042$.

The set of available assets, that is the set of assets considered for an investment, is denoted by $N = \{1, 2, \ldots, n\}$. The rate of return at the target time of each asset j, $j = 1, \ldots, n$, is modeled as a random variable R_j with given mean $\mu_j = \mathbb{E}\{R_j\}$.

Let x_j denote the decision variable expressing the weight of asset j, $j = 1, \ldots, n$. Let $\mathbf{x} = (x_j)_{j=1,\ldots,n}$ denote a vector of decision variables x_j. We will also say that $\mathbf{x}$ is a portfolio. In order to represent a portfolio the weights must satisfy a set of constraints. A basic requirement is that the weights must sum to one, i.e.

$$\sum_{j=1}^{n} x_j = 1. \tag{1.1}$$

Moreover, we assume that it is not allowed to buy assets without the necessary capital, that is *short sales* are not allowed. Thus, we impose

$$x_j \geq 0 \qquad j = 1, \ldots, n. \tag{1.2}$$

An investor may wish to consider other requirements on the weights. For example, minimum or maximum values on the weights of some specific assets may be set. The constraint on a maximum weight of 30 % for asset j would be simply formalized as $x_j \leq 0.3$. Also, a minimum or a maximum bound may be set on the total weight of a set of assets (for example, the assets of a certain sector). The constraint on a total weight not greater than 40 % on assets i and j would be $x_i + x_j \leq 0.4$. In general, the weights are subject to a set of additional side constraints. Most of them can be expressed as linear equations and inequalities. We denote with Q the set of feasible portfolios, that is the set of portfolios that satisfy the whole system of linear equations and inequalities, including (1.1) and (1.2). Then, we will write

$$\mathbf{x} \in Q.$$

There may be requirements on the $\mathbf{x}$ that cannot be expressed through linear equation or inequalities. In Chaps. 3 and 4 real features such as transaction costs, minimum transaction lots, cardinality constraints, thresholds on maximum or minimum investments will be discussed and modeled in most cases through the introduction of integer and binary variables.

Each portfolio $\mathbf{x}$ is associated with a random variable $R_\mathbf{x} = \sum_{j=1}^{n} R_j x_j$ that represents the portfolio rate of return and is the weighted sum of the rates of return

1.6 Markowitz Model

of the assets. The mean rate of return of portfolio **x** is given as:

$$\mu(\mathbf{x}) = \mathbb{E}\{R_{\mathbf{x}}\} = \sum_{j=1}^{n} \mu_j x_j.$$

Maximizing the expected rate of return of the portfolio is one of the goals of an investor. The other goal is to protect the investment from the uncertainty. We now introduce the concept of risk formally. We indicate with $\varrho(\mathbf{x})$ a measure of risk associated with portfolio **x**. For the time being, we do not specify the functional structure of $\varrho(\mathbf{x})$. We focus on the class of risk measures that are equal to 0 in the case of a risk-free portfolio, that is a portfolio whose rate of return is known with certainty, and take positive values for any other portfolio. Such measures are called *shift independent dispersion measures*.

In its high level formulation, portfolio optimization is modeled as a mean-risk bi-criteria optimization problem:

$$\max\{[\mu(\mathbf{x}), -\varrho(\mathbf{x})] : \quad \mathbf{x} \in Q\}, \tag{1.3}$$

where the mean rate of return $\mu(\mathbf{x})$ of the portfolio is maximized and the risk measure $\varrho(\mathbf{x})$ is minimized. This is a conceptual formulation as no portfolio will at the same time maximize the mean return and minimize the risk. Typically, a portfolio with high mean return will be highly risky, and viceversa. A feasible portfolio $\mathbf{x}^0 \in Q$ is called an efficient solution of problem (1.3) or a μ/ϱ-*efficient* portfolio if there is no $\mathbf{x} \in Q$ such that $\mu(\mathbf{x}) \geq \mu(\mathbf{x}^0)$ and $\varrho(\mathbf{x}) \leq \varrho(\mathbf{x}^0)$ with at least one inequality strict. Note that in Fig. 1.5 not all portfolios of the risk/return frontier are μ/ϱ-efficient.

The relation 'not less efficient than' defines a partial ordering on any set of feasible portfolios. The concept of efficient portfolios has a key role in portfolio optimization. Any efficient portfolio may be interesting for a class of investors. Investors with high risk aversion will tend to prefer efficient portfolios with low risk, and thus usually with low expected return. Viceversa, risk seekers will prefer portfolios with high expected return, and thus high risk. On the other hand, no rational investor should accept a portfolio that is not μ/ϱ-efficient.

1.6 Markowitz Model

The expression *Modern Portfolio Theory* (MPT) is widely used to identify the optimization theory and models that aim at optimizing the investment in assets, typically by maximizing the portfolio expected return for a given level of portfolio risk, or by minimizing the risk for a given level of expected return. The origin of MPT dates back to the work carried out in the fifties by Harry Markowitz, who later won the Nobel prize for his scientific contributions. For the first time Markowitz mathematically formalized the concepts of expected return and risk, structured the concept of diversification, and formulated an optimization model. In his earth-breaking work, he suggested to use the variance

$$\sigma^2 = \mathbb{E}\{(R - \mathbb{E}\{R\})^2\} \tag{1.4}$$

as a risk measure and wrote that 'the returns from assets are too inter-correlated. Diversification cannot eliminate all variance'. For a given expected rate of return, the variance of portfolio return can be reduced through diversification, but cannot be eliminated. The variance can be reduced further only by decreasing the expected portfolio rate of return.

We now show that the variance of portfolio return can be expressed as a quadratic function of the weights x_j of the assets. We recall that the rate of return R_j of asset j is a random variable with mean μ_j. We denote by σ_j^2 the variance of R_j and by σ_{ij} the covariance between the rates of return of assets i and j, calculated as

$$\sigma_{ij} = \mathbb{E}\{(R_i - \mu_i)(R_j - \mu_j)\}.$$

When normalized, the covariance leads to the correlation coefficients $\bar{\rho}_{ij}$ calculated as

$$\bar{\rho}_{ij} = \frac{\sigma_{ij}}{\sigma_i \sigma_j}. \tag{1.5}$$

A positive value of the correlation coefficient indicates that the rates of return of assets i and j tend to move in the same direction, upward or downward. The larger the value of the coefficient is, the stronger the tendency is. Pairs of assets belonging to the same commercial sector, e.g. automotive or bank, typically have positive correlation coefficients. A negative value indicates that the rates of return tend to move in opposite directions, that is when one increases the other one tends to decrease.

The variance of the rate of return $R_\mathbf{x}$ of portfolio $\mathbf{x}$ can be expressed as

$$\sigma^2(\mathbf{x}) = \sum_{i=1}^{n} \sigma_i^2 x_i^2 + 2 \sum_{i=1}^{n} \sum_{j=i+1}^{n} \sigma_{ij} x_i x_j. \tag{1.6}$$

An equivalent expression can be obtained by denoting the variance of asset i as σ_{ii}:

$$\sigma^2(\mathbf{x}) = \sum_{i=1}^{n} \sum_{j=1}^{n} \sigma_{ij} x_i x_j.$$

1.6 Markowitz Model

In Markowitz model the variance is minimized and a lower bound μ_0 is set on the expected portfolio rate of return. The classical form of Markowitz model is:

$$\min \sum_{i=1}^{n} \sum_{j=1}^{n} \sigma_{ij} x_i x_j \tag{1.7a}$$

$$\sum_{j=1}^{n} \mu_j x_j \geq \mu_0 \tag{1.7b}$$

$$\sum_{j=1}^{n} x_j = 1 \tag{1.7c}$$

$$x_j \geq 0 \quad j = 1, \ldots, n. \tag{1.7d}$$

This is a quadratic programming problem with quadratic objective function, two linear constraints and non-negative continuous variables. The problem is simple, but already very interesting.

We now show that diversification reduces the variance of the portfolio rate of return and observe the impact of the correlation coefficient by means of an example with two assets, A and B. From (1.6) and (1.5) the variance of a portfolio composed of A and B is:

$$\sigma^2(\mathbf{x}) = \sigma_A^2 x_A^2 + \sigma_B^2 x_B^2 + 2\bar{\rho}_{AB} \sigma_A \sigma_B x_A x_B. \tag{1.8}$$

The variance depends on the individual variances of the asset rates of return and on their correlation. For given values of x_A and x_B Eq. (1.8) shows that the variance of the portfolio rate of return decreases when decreasing $\bar{\rho}_{AB}$ from 1 to -1.

In Fig. 1.6 we draw five risk/return frontiers obtained by setting the correlation coefficient $\bar{\rho}_{AB}$ to $-1, -0.5, 0, 0.5, 1$, respectively. The standard deviation of assets A and B is $\sigma_A = 8\%$ and $\sigma_B = 18\%$, whereas the expected returns are $\mu_A = 3\%$ and $\mu_B = 14\%$. The figure shows that, for a given level of expected return, the standard deviation of an efficient portfolio decreases when the correlation coefficient decreases from 1 to -1.

Let us analyze the extreme cases:

1. $\bar{\rho}_{AB} = 1$: the assets are perfectly positively correlated. In this case the frontier is a straight line between the points associated with the portfolios composed by one asset only. This can be seen also from (1.8):

$$\sigma^2(\mathbf{x}) = \sigma_A^2 x_A^2 + \sigma_B^2 x_B^2 + 2\sigma_A \sigma_B x_A x_B = (\sigma_A x_A + \sigma_B x_B)^2.$$

Then, the portfolio standard deviation is the convex combination of the standard deviations of the two assets, with coefficients x_A and x_B, $x_A + x_B = 1$.

2. $\bar{\rho}_{AB} = -1$: assets are perfectly negatively correlated. The frontier consists of two lines intersecting on the expected return axis (null standard deviation). This

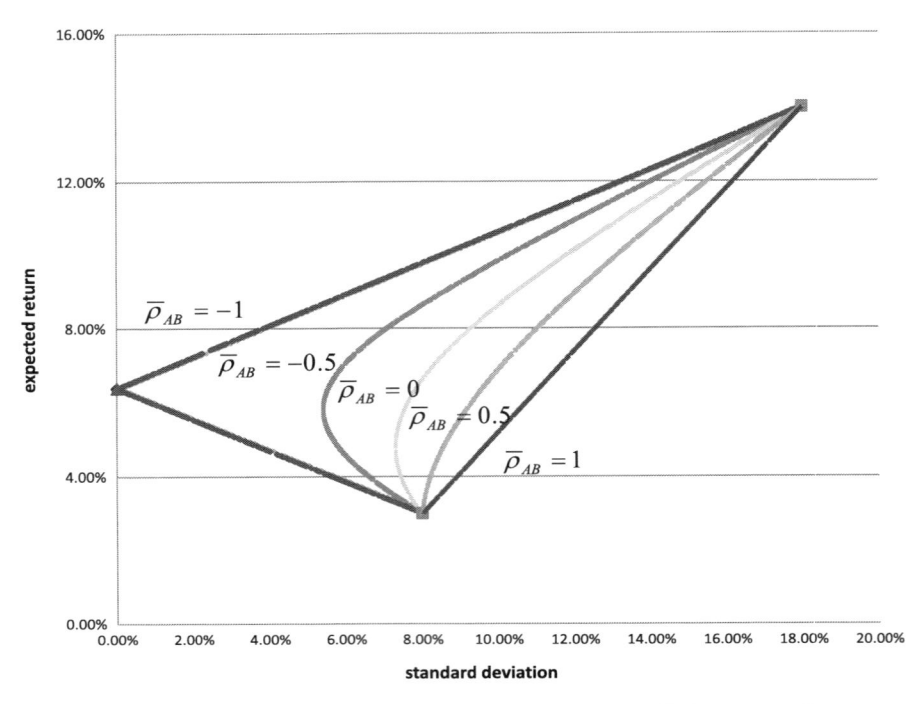

Fig. 1.6 Markowitz risk/return frontiers with different correlation coefficients: case of two assets

behavior can be explained again by analyzing formula (1.8). In this case the equation becomes

$$\sigma^2(\mathbf{x}) = \sigma_A^2 x_A^2 + \sigma_B^2 x_B^2 - 2\sigma_A \sigma_B x_A x_B = (\sigma_A x_A - \sigma_B x_B)^2.$$

Since $x_B = 1 - x_A$, the standard deviation $\sigma_A x_A - \sigma_B x_B$ becomes null when $x_A = \frac{\sigma_B}{\sigma_A + \sigma_B}$. For values of x_A lower and greater than $\frac{\sigma_B}{\sigma_A + \sigma_B}$ two straight lines are obtained.

1.7 Risk and Safety Measures

The risk measures we consider in this book are dispersion measures that quantify the level of variability of the portfolio rate of return around its expected value. This looks like a sensible way to measure the variability, but has a drawback. A portfolio with extremely low return is efficient if such return can be achieved with certainty, that is if the dispersion is null. Such portfolio is not dominated by any other risky portfolio, even when all the possible values of the rate of return are larger. This can be illustrated by the following example.

1.7 Risk and Safety Measures

Example 1.2 Let us consider two portfolios $\mathbf{x}'$ and $\mathbf{x}''$ with rates of return given in percentages:

$$\mathbb{P}\{R_{\mathbf{x}'} = \xi\} = \begin{cases} 1, & \xi = 1\,\% \\ 0, & \text{otherwise} \end{cases} \quad \mathbb{P}\{R_{\mathbf{x}''} = \xi\} = \begin{cases} 1/2, & \xi = 3\,\% \\ 1/2, & \xi = 5\,\% \\ 0, & \text{otherwise}. \end{cases}$$

The two portfolios are both efficient for any dispersion measure, as $\mathbf{x}'$ has lower return than $\mathbf{x}''$ but is a risk-free portfolio. However, $\mathbf{x}'$ with the guaranteed return 1 % is obviously worse than the risky portfolio $\mathbf{x}''$ whose return may be either 3 % or 5 %. No rational investor would prefer $\mathbf{x}'$ to $\mathbf{x}''$, because under any possible realization of the rate of return, portfolio $\mathbf{x}''$ performs better than $\mathbf{x}'$.

In order to overcome this weakness of the dispersion measures, the concept of safety measure, that is a measure that an investor aims at maximizing, was introduced.

Some of the measures that will be presented in Chap. 2 are risk measures, that is dispersion measures, and some are safety measures which, when embedded in an optimization model, are maximized instead of being minimized. Each risk measure $\varrho(\mathbf{x})$ has a well defined corresponding safety measure $\mu(\mathbf{x}) - \varrho(\mathbf{x})$ and viceversa.

In a high level formulation, the portfolio optimization problem is modeled as a mean-safety bi-criteria optimization problem:

$$\max\{[\mu(\mathbf{x}), \mu(\mathbf{x}) - \varrho(\mathbf{x})] : \quad \mathbf{x} \in Q\}. \tag{1.9}$$

Although the risk measures are more 'natural', due to the consolidated familiarity with Markowitz model (see Sect. 1.6), the safety measures do not suffer of the drawback we observed for the dispersion measures. The concepts will be formalized in Chap. 2.

A portfolio dominated in the mean-risk problem (1.3) can be shown to be dominated also in the corresponding mean-safety problem (1.9). Indeed, if portfolio $\mathbf{x}'$ is dominated by $\mathbf{x}''$ in the mean-risk problem, then $\mu(\mathbf{x}'') \geq \mu(\mathbf{x}')$ and $\varrho(\mathbf{x}'') \leq \varrho(\mathbf{x}')$ with at least one inequality strict. From this, we derive that

$$\varrho(\mathbf{x}'') - \mu(\mathbf{x}'') < \varrho(\mathbf{x}') - \mu(\mathbf{x}').$$

Thus, portfolio $\mathbf{x}'$ is dominated by portfolio $\mathbf{x}''$ also in the mean-safety problem. Hence, the efficient portfolios of problem (1.9) form a subset of the entire μ/ϱ-efficient set. We illustrate this in Fig. 1.7. As we will see in Chap. 6, the risk measure $\varrho(\mathbf{x})$ is typically a convex function. Due to linearity of $\mu(\mathbf{x})$ and convexity of $\varrho(\mathbf{x})$, the portfolios $\mathbf{x} \in Q$ form a set with the convex boundary from the side of μ-axis (i.e., the set $\{(\mu, \varrho) : \mu = \mu(\mathbf{x}), \varrho \geq \varrho(\mathbf{x}), \mathbf{x} \in Q\}$ is convex). This boundary represents a curve of the minimum risk portfolios spanning from the best expectation portfolio (BEP) to the worst expectation portfolio (WEP). The

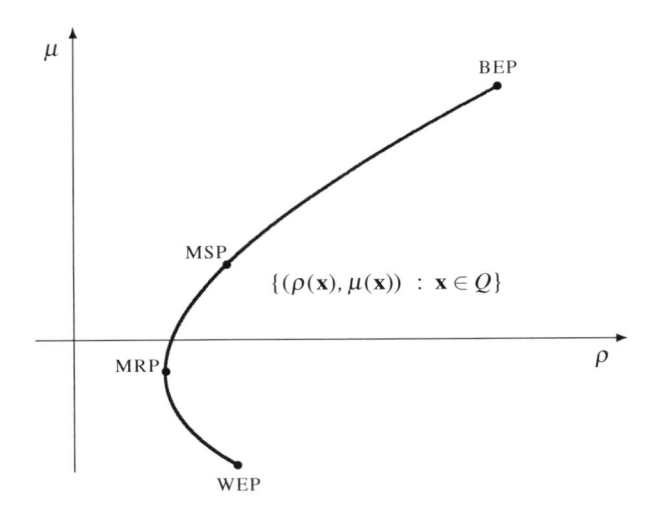

Fig. 1.7 The mean-risk and mean-safety analysis

Minimum Risk Portfolio (MRP), defined as the solution of $\min_{\mathbf{x} \in Q} \varrho(\mathbf{x})$, limits the curve to the mean-risk efficient frontier from BEP to MRP. Similarly, the Maximum Safety Portfolio (MSP), defined as the solution of $\max_{\mathbf{x} \in Q}\{\mu(\mathbf{x}) - \varrho(\mathbf{x})\}$, distinguishes a part of the mean-risk efficient frontier, from BEP to MSP, which is also mean-safety efficient.

For specific types of return distributions or specific feasible sets, the subset of rational portfolios may exceed the limit of the MSP. Hence, the mean-safety problem (1.9) may be too restrictive in such situations and it may be important to keep the full modeling capabilities of the original mean-risk approach (1.3).

1.8 Handling Bi-Criteria Optimization Problems

In the previous section the bi-criteria mean-risk and mean-safety problems were formulated. We first consider the mean-risk problem (1.3). There are two approaches to handle it: the bounding approach and the trade-off analysis. The former is a common approach based on the use of a specified lower bound μ_0 on the expected return of the portfolio which results in the following problem:

$$\min\{\varrho(\mathbf{x}): \quad \mu(\mathbf{x}) \geq \mu_0, \quad \mathbf{x} \in Q\}. \tag{1.10}$$

1.8 Handling Bi-Criteria Optimization Problems

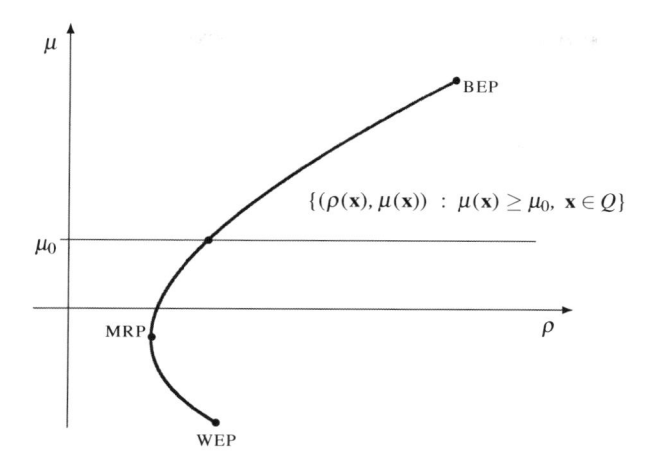

Fig. 1.8 The bounding approach

This bounding approach provides a clear understanding of investor preferences. One may also use a bounding approach where the risk is bounded:

$$\max\{\mu(\mathbf{x}) : \quad \varrho(\mathbf{x}) \leq \varrho_0, \quad \mathbf{x} \in Q\}. \tag{1.11}$$

Due to the convexity of risk measures $\varrho(\mathbf{x})$ with respect to $\mathbf{x}$, by solving the parametric problem (1.10) with changing $\mu_0 \in [\min_{j=1,\ldots,n} \mu_j, \max_{j=1,\ldots,n} \mu_j]$ various efficient portfolios are obtained. Actually, for μ_0 smaller than the expected return of the MRP, problem (1.10) generates always the MRP as optimal solution. Larger values of μ_0 provide the parameterization of the μ/ϱ–efficient frontier by generating efficient portfolios with $\mu(\mathbf{x}) = \mu_0$ (see Fig. 1.8).

The approach (1.10) generates also those portfolios that are efficient solutions of the corresponding mean–safety problem (1.9). Portfolios corresponding to values of μ_0 exceeding the expected return of the MSP are also efficient solutions to the corresponding mean-safety problem (1.9). However, having a specified value of parameter μ_0 it is not known if the optimal solution of (1.10) is also an efficient portfolio with respect to the corresponding mean–safety problem (1.9) or not. Therefore, when using the bounding approach a separate problem

$$\max\{\mu(\mathbf{x}) - \varrho(\mathbf{x}) : \quad \mu(\mathbf{x}) \geq \mu_0, \quad \mathbf{x} \in Q\} \tag{1.12}$$

should be considered explicitly for the corresponding mean-safety problem (1.9).

Another approach to the bi-criteria mean-risk problem is to take advantage of the efficient frontier convexity and perform a trade-off analysis. Assuming a trade-off coefficient λ between the risk and the return, the so-called *risk aversion coefficient*,

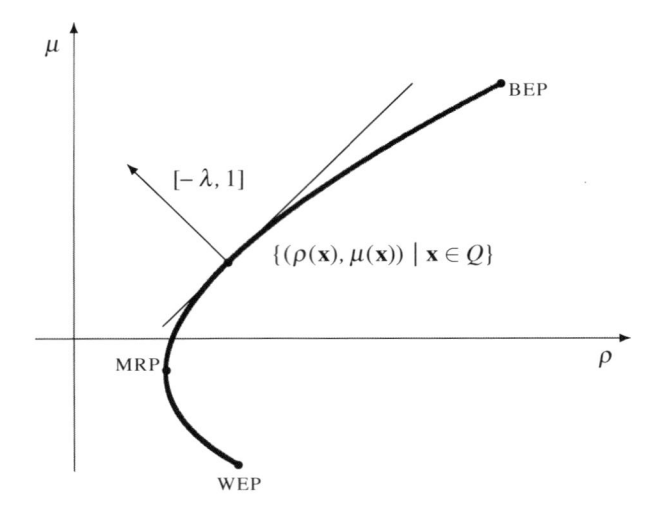

Fig. 1.9 The trade-off approach

the best portfolio can be found by solving the *trade-off optimization problem*:

$$\max \{\mu(\mathbf{x}) - \lambda \varrho(\mathbf{x}) : \quad \mathbf{x} \in Q\}. \tag{1.13}$$

Different positive values of parameter λ allow the generation of different efficient portfolios. The so-called *critical line approach* is obtained by solving the parametric problem (1.13) with changing $\lambda > 0$. Due to convexity of risk measures $\varrho(\mathbf{x})$ with respect to $\mathbf{x}$, the values $\lambda > 0$ provide the parameterization of the entire set of the μ/ϱ–efficient portfolios (with the exception of its two ends BEP and MRP which are the limiting cases). Note that $(1 - \lambda)\mu(\mathbf{x}) + \lambda(\mu(\mathbf{x}) - \varrho(\mathbf{x})) = \mu(\mathbf{x}) - \lambda\varrho(\mathbf{x})$. Hence, bounded trade-off $0 < \lambda < 1$ in the mean-risk problem (1.3) corresponds to the complete weighting parameterization of the mean-safety problem (1.9) (see Fig. 1.9). Contrary to the bounding approach, having a specified value of parameter λ makes it possible to immediately know if the optimal solution of (1.13) is also an efficient portfolio for the mean-safety problem (1.9) or not.

Thus, the trade-off problem (1.13) offers a tool for both the mean-risk and the corresponding mean-safety problems and provides easy modeling of the risk aversion by means of the trade-off value λ.

An alternative specific approach looks for a risky portfolio offering the maximum increase of the mean return, compared to the risk-free investment opportunities. Namely, given the risk-free rate of return r_0, a risky portfolio $\mathbf{x}$ that maximizes the ratio $(\mu(\mathbf{x}) - r_0)/\varrho(\mathbf{x})$ is sought. This leads us to the following ratio optimization problem:

$$\max \left\{ \frac{\mu(\mathbf{x}) - r_0}{\varrho(\mathbf{x})} : \mathbf{x} \in Q \right\}. \tag{1.14}$$

1.9 Notes and References

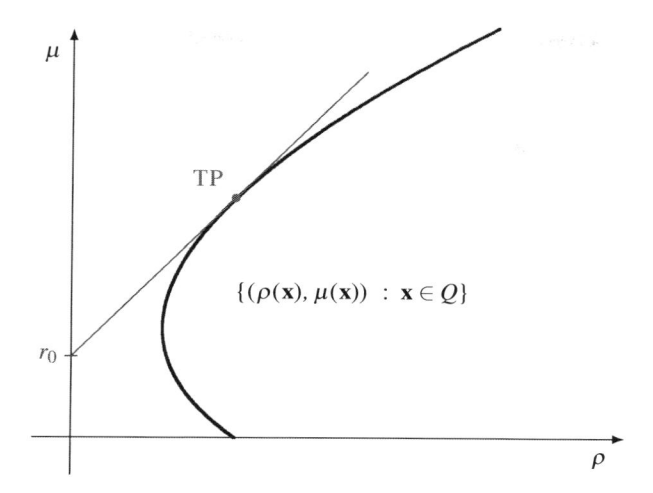

Fig. 1.10 The tangency portfolio

The optimal solution of problem (1.14) is usually called the *tangency portfolio* (TP) or the *market portfolio*. We illustrate it in Fig. 1.10. The so-called Capital Market Line (CML) is the line drawn from the intercept corresponding to r_0 and that passes tangent to the risk/return frontier. Any point on this line provides the maximum return for each level of risk. The tangency portfolio is the portfolio of risky assets corresponding to the point where the CML is tangent to the risk/return frontier.

1.9 Notes and References

Harry Markowitz is considered the father of Modern Portfolio Theory. His pioneering work in 1952 (Markowitz 1952) represents the first contribution in terms of an optimization model for portfolio selection seeking return maximization given a level of risk or risk minimization given a level of return. The concept of efficient frontier was developed in Markowitz (1959). Afterward, in a historical contribution Sharpe (1964) proposed the Capital Asset Pricing Model (CAPM), a theory of market equilibrium, further developed independently by Lintner (1965) and Mossin (1966). The CAPM establishes a relation between the return of an asset and a single risk factor that considers, besides the asset risk, also the market risk. For this contribution William Sharpe, together with Merton Miller and Harry Markowitz, won the Nobel prize in economic sciences in 1990. From this starting point, several multi-factor risk models have been developed. Interested readers are referred to Elton and Gruber (1995) and Elton et al. (2003).

Baumol (1964) was the first to suggest a safety measure, which he called the expected gain-confidence limit criterion. The ratio optimization model (1.14) follows the classical model of Tobin (1958) for portfolio leverage, where the Capital Market Line is drawn in the risk/return space with risk measured by standard deviation.

In the last years both academic researchers and practitioners have been discussing about the advantages and drawbacks of Markowitz model, in particular on the realism of the assumptions upon which the model is based. Some of the criticisms are: investors may not be interested in the model, because the model does not capture their utility function; the model assumes that asset returns have a symmetric distribution; correlations between assets are fixed and constant; there are no taxes or transaction costs; all assets can be divided into parcels of any size; risk of an asset is known in advance and is constant. Markowitz (1952) himself wrote 'It is a story of which I have read only the first page of the first chapter.' In the rest of this book, we will see further models that overcome several of the criticisms.

Chapter 2
Linear Models for Portfolio Optimization

2.1 Introduction

Nowadays, Quadratic Programming (QP) models, like Markowitz model, are not hard to solve, thanks to technological and algorithmic progress. Nevertheless, Linear Programming (LP) models remain much more attractive from a computational point of view for several reasons. The design and development of commercial software for the solution of LP models is more advanced than for QP models. As a consequence, several commercial LP solvers are available and, in general, LP solvers tend to be more reliable than QP solvers. On average, LP solvers can solve in small time (the order of seconds) instances of much larger size than QP solvers.

Is it possible to have linear models for portfolio optimization? How can we measure the risk or safety in order to have a linear model? A first observation is that, in order to guarantee that a portfolio takes advantage of diversification, no risk or safety measure can be a linear function of the shares of the assets in the portfolio, that is of the variables x_j, $j = 1, \ldots, n$. Linear models, however, can be obtained through discretization of the return random variables or, equivalently, through the concept of scenarios.

2.2 Scenarios and LP Computability

We have indicated by R_j the random variable representing the rate of return of asset j, $j = 1, \ldots, n$, at the target time.

Now we change the way we look at the uncertainty of the rates of return of the assets at the target time and introduce the concept of *scenario*. A scenario is, informally, a possible situation that can happen at the target time, in our case a possible realization of the rates of return of the assets at the target time. Depending

Table 2.1 Scenarios: An example

Asset	Scenario 1 (%)	Scenario 2 (%)	Scenario 3 (%)	Mean return rate (%)
1	$r_{11} = 3.1$	$r_{12} = -2.7$	$r_{13} = 1.60$	$\mu_1 = 0.67$
2	$r_{21} = 2.3$	$r_{22} = -2.3$	$r_{23} = 1.30$	$\mu_2 = 0.43$
3	$r_{31} = 4.2$	$r_{32} = -3.1$	$r_{33} = -0.2$	$\mu_3 = 0.43$
4	$r_{41} = 1.5$	$r_{42} = -2.0$	$r_{43} = -0.1$	$\mu_4 = -0.2$

on what will happen between the investment time and the target time, any of several different scenarios may become true. The scenarios may also be less or more likely to happen. More formally, a scenario is a realization of the multivariate random variable representing the rates of return of all the assets.

We now suppose that, on the basis of a careful preliminary analysis, T different scenarios have been identified as possible at the target time. The probability that scenario t, $t = 1, \ldots, T$, will happen is indicated by p_t, with $\sum_{t=1}^{T} p_t = 1$. We assume that for each random variable R_j, $j = 1, \ldots, n$, its realization r_{jt} under scenario t is known. The set of the returns of all the assets $\{r_{jt}, j = 1, \ldots, n\}$ defines the scenario t. The expected return of asset j, $j = 1, \ldots, n$, is calculated as $\mu_j = \sum_{t=1}^{T} p_t r_{jt}$. The concept of scenario captures the correlation among the rates of return of the assets.

In Table 2.1, we show an example of $n = 4$ assets and $T = 3$ scenarios. The table shows the rates of return of the assets in the different scenarios. Scenario 1 is an optimistic scenario: all rates of return are positive. Scenario 2 is negative, whereas scenario 3 is positive for assets 1 and 2 and negative for assets 3 and 4. The averages are computed under the assumptions that the scenarios are equally probable ($p_t = 1/3, t = 1, 2, 3$).

Identifying the scenarios, their probabilities and estimating the values of the rate of return r_{jt} of each asset j under each scenario t is crucial. To be statistically significant, the number of scenarios has to be sufficiently large.

Each portfolio $\mathbf{x}$ defines a corresponding random variable $R_\mathbf{x} = \sum_{j=1}^{n} R_j x_j$ that represents the portfolio rate of return. The step-wise cumulative distribution function (cdf) of $\{R_\mathbf{x}\}$ is defined as

$$F_\mathbf{x}(\xi) = P(R_\mathbf{x} \leq \xi). \tag{2.1}$$

The return y_t of a portfolio $\mathbf{x}$ in scenario t can be computed as

$$y_t = \sum_{j=1}^{n} r_{jt} x_j, \tag{2.2}$$

and the expected return of the portfolio $\mu(\mathbf{x})$ can be computed as a linear function of $\mathbf{x}$

$$\mu(\mathbf{x}) = \mathbb{E}\{R_\mathbf{x}\} = \sum_{t=1}^{T} p_t y_t = \sum_{t=1}^{T} p_t (\sum_{j=1}^{n} r_{jt} x_j) = \sum_{j=1}^{n} x_j \sum_{t=1}^{T} p_t r_{jt} = \sum_{j=1}^{n} \mu_j x_j. \tag{2.3}$$

We have defined a scenario as a realization of the multivariate random variable representing the rates of return of the assets. We may look at the set of scenarios as a discretization of the multivariate random variable.

We will say that the returns are *discretized* when they are defined by their realizations under the specified scenarios, that is by the set of values $\{r_{jt} : j = 1, \ldots, n, t = 1, \ldots, T\}$. We will say that a risk or a safety measure is *LP computable* if the portfolio optimization model takes a linear form in the case of discretized returns.

2.3 Basic LP Computable Risk Measures

The variance is the classical statistical quantity used to measure the dispersion of a random variable around its mean. There are, however, other ways to measure the dispersion of a random variable. The random variable, we are interested in, is the portfolio return $R_\mathbf{x}$.

The Mean Absolute Deviation (MAD) is a dispersion measure that is defined as

$$\delta(\mathbf{x}) = \mathbb{E}\{|R_\mathbf{x} - \mathbb{E}\{R_\mathbf{x}\}|\} = \mathbb{E}\{|\sum_{j=1}^{n} R_j x_j - \mathbb{E}\{\sum_{j=1}^{n} R_j x_j\}|\}. \tag{2.4}$$

The MAD measures the average of the absolute value of the difference between the random variable and its expected value. With respect to the variance, the MAD considers absolute values instead of squared values. We show in the following that, when the returns are discretized, the MAD is LP computable. Recalling that the expected return of the portfolio can be calculated as (2.3), the MAD can be written as

$$\delta(\mathbf{x}) = \sum_{t=1}^{T} p_t (|\sum_{j=1}^{n} r_{jt} x_j - \sum_{j=1}^{n} \mu_j x_j|). \tag{2.5}$$

The portfolio optimization problem then becomes

$$\min \delta(\mathbf{x}) = \sum_{t=1}^{T} p_t (|\sum_{j=1}^{n} r_{jt}x_j - \mu|) \quad (2.6a)$$

$$\mu = \sum_{j=1}^{n} \mu_j x_j \quad (2.6b)$$

$$\mu \geq \mu_0 \quad (2.6c)$$

$$\mathbf{x} \in Q, \quad (2.6d)$$

where μ_0 is the lower bound on the portfolio expected return required by the investor, and Q denotes the system of constraints defining the set of feasible portfolios as described in Chap. 1.

This form is not linear in the variables x_j but can be transformed into a linear form. Using (2.2) for the return of the portfolio in scenario t, y_t, $\delta(\mathbf{x})$ can also be written as

$$\delta(\mathbf{x}) = \sum_{t=1}^{T} p_t (|y_t - \sum_{j=1}^{n} \mu_j x_j|).$$

We now define the deviation in scenario t as d_t, that is $d_t = |y_t - \sum_{j=1}^{n} \mu_j x_j|$. Then, the portfolio optimization problem is

$$\min \sum_{t=1}^{T} p_t d_t \quad (2.7a)$$

$$d_t = |y_t - \mu| \qquad t = 1, \ldots, T \quad (2.7b)$$

$$y_t = \sum_{j=1}^{n} r_{jt} x_j \qquad t = 1, \ldots, T \quad (2.7c)$$

$$\mu = \sum_{j=1}^{n} \mu_j x_j \quad (2.7d)$$

$$\mu \geq \mu_0 \quad (2.7e)$$

$$\mathbf{x} \in Q. \quad (2.7f)$$

2.3 Basic LP Computable Risk Measures

Since $|y_t - \mu| = \max\{(y_t - \mu); -(y_t - \mu)\}$, the problem can be written in the following equivalent linear form

$$\min \sum_{t=1}^{T} p_t d_t \tag{2.8a}$$

$$d_t \geq y_t - \mu \qquad t = 1, \ldots, T \tag{2.8b}$$

$$d_t \geq -(y_t - \mu) \qquad t = 1, \ldots, T \tag{2.8c}$$

$$y_t = \sum_{j=1}^{n} r_{jt} x_j \qquad t = 1, \ldots, T \tag{2.8d}$$

$$\mu = \sum_{j=1}^{n} \mu_j x_j \tag{2.8e}$$

$$\mu \geq \mu_0 \tag{2.8f}$$

$$d_t \geq 0 \qquad t = 1, \ldots, T \tag{2.8g}$$

$$\mathbf{x} \in Q. \tag{2.8h}$$

The equivalence comes from observing that if $y_t - \mu \geq 0$ constraints (2.8c) are redundant. In this case constraints (2.8b), combined with the minimization of $\sum_{t=1}^{T} p_t d_t$ in (2.8a) that pushes the value of each d_t to the minimum value allowed by the constraints, impose that $d_t = y_t - \mu = |y_t - \mu|$. If, on the contrary $y_t - \sum_{j=1}^{n} \mu_j x_j \leq 0$, constraints (2.8b) are redundant. In this case, constraints (2.8c), combined with the objective function, impose that $d_t = -(y_t - \sum_{j=1}^{n} \mu_j x_j) = |y_t - \sum_{j=1}^{n} \mu_j x_j|$. Thus, in conclusion, the optimization model (2.8) is a linear programming model for the optimization of a portfolio where the risk is measured through the MAD of the return of the portfolio.

In Fig. 2.1, we represent the calculation of the MAD measure. In other words, we assume that the values of the shares x_j are given. We represent over the horizontal axis the scenarios $t = 1, \ldots, T$ and over the vertical axis the values y_t of the return

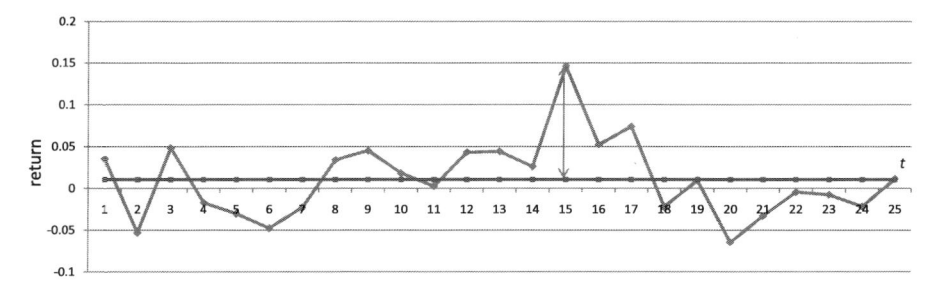

Fig. 2.1 The MAD measure

of the portfolio under the various scenarios t. The thick horizontal line identifies the expected return of the portfolio $\mu = \sum_{j=1}^{n} \mu_j x_j = \sum_{t=1}^{T} p_t y_t$. The length of a vertical segment is the absolute value of the deviation d_t (in Fig. 2.1, the deviation d_{15} corresponding to the scenario $t = 15$ is drawn as example). The MAD model aims at minimizing the average absolute deviation.

In the case the rates of return are a multivariate normally distributed random variable, the rate of return of the portfolio is normally distributed. Then, the proportionality relation between the mean absolute deviation and the standard deviation occurs $\delta(\mathbf{x}) = \sqrt{\frac{2}{\pi}} \sigma(\mathbf{x})$. As a consequence, minimizing the MAD is equivalent to minimizing the standard deviation, which means, in this specific case, the equivalence of the associated optimization problems. However, the MAD model does not require any specification of the return distribution.

The MAD accounts for all deviations of the rate of return of the portfolio from its expected value, both below and above the expected value. However, one may sensibly think that any rational investor would consider real risk only the deviations below the expected value. In other words, the variability of the portfolio rate of return above the mean should not be penalized since the investors are concerned with under-performance rather than over-performance of a portfolio. In terms of scenarios, the risky scenarios are those where the rate of return of the portfolio is below its expected value. We can modify the definition of the MAD in order to consider only the deviations below the expected value. We define the Semi Mean Absolute Deviation (Semi-MAD)

$$\bar{\delta}(\mathbf{x}) = \mathbb{E}\{\max\{0, \mathbb{E}\{\sum_{j=1}^{n} R_j x_j\} - \sum_{j=1}^{n} R_j x_j\}\}, \quad (2.9)$$

where the deviations above the expected value are not calculated. The portfolio optimization problem (2.8) presented for the MAD can be adapted to the Semi-MAD as follows:

$$\min \sum_{t=1}^{T} p_t d_t \qquad (2.10\text{a})$$

$$d_t \geq \mu - y_t \qquad t = 1, \ldots, T \qquad (2.10\text{b})$$

$$y_t = \sum_{j=1}^{n} r_{jt} x_j \qquad t = 1, \ldots, T \qquad (2.10\text{c})$$

$$\mu = \sum_{j=1}^{n} \mu_j x_j \qquad (2.10\text{d})$$

$$\mu \geq \mu_0 \qquad (2.10\text{e})$$

$$d_t \geq 0 \qquad t = 1, \ldots, T \qquad (2.10\text{f})$$

$$\mathbf{x} \in Q. \qquad (2.10\text{g})$$

2.3 Basic LP Computable Risk Measures

The formulation for the Semi-MAD is the formulation of the MAD, from which inequalities (2.8b) have been dropped. If, for a given scenario t, $\mu - y_t > 0$, this means that under scenario t the rate of return of the portfolio y_t is below the expected value. In this case d_t in the optimum will be the difference $\mu - y_t$. If instead $\mu - y_t \leq 0$, constraint (2.10b) becomes redundant and in the optimum $d_t = 0$. Thus, the deviations above the expected value are not calculated in the objective function.

The Semi-MAD seems to be a very attractive measure, focusing on the downside risk only. However, it can be seen that it is equivalent to the MAD as the corresponding optimization models generate the same optimal portfolio. The intuition behind the equivalence, that is somehow surprising, is that the MAD is the sum of the deviations above and below the expected value. By definition of expected value, the sum of the deviations above the expected value is equal to the sum of the deviations below the expected value. Thus, the Semi-MAD is half the MAD. Minimizing the downside deviations is equivalent to minimizing the total deviations and equivalent to minimizing the deviations above the expected value as well. We make this equivalence formal.

Theorem 2.1 *Minimizing the MAD is equivalent to minimizing the Semi-MAD as $\delta(\mathbf{x}) = 2\bar{\delta}(\mathbf{x})$.*

Proof We first write the mean deviation of the portfolio rate of the return from its expected value and show that it is equal to 0:

$$\mathbb{E}\{R_\mathbf{x} - \mathbb{E}\{R_\mathbf{x}\}\} = \mathbb{E}\{R_\mathbf{x}\} - \mathbb{E}\{R_\mathbf{x}\} = 0$$

From this it immediately follows that the average positive deviation ($y_t - \mu(\mathbf{x}) > 0$ implies the rate of return of the portfolio in scenario t is above its expected value) is equal to the opposite of the average negative deviation ($y_t - \mu(\mathbf{x}) < 0$ implies the rate of return of the portfolio in scenario t is below its expected value). The absolute value of the average positive deviation is thus equal to the absolute value of average negative deviation, from which it follows that the MAD is twice the Semi-MAD.

□

Although the MAD has become a very popular risk measure, a different LP computable risk measure was earlier proposed, namely the Gini's mean difference. The variability of the portfolio return is captured here by the differences of the portfolio returns in different scenarios. For a discrete random variable represented by its realizations y_t, the *Gini's mean difference (GMD)* considers as risk the average absolute value of the differences of the portfolio returns y_t in different scenarios:

$$\Gamma(\mathbf{x}) = \frac{1}{2} \sum_{t'=1}^{T} \sum_{t''=1}^{T} |y_{t'} - y_{t''}| p_{t'} p_{t''}. \tag{2.11}$$

The risk function $\Gamma(\mathbf{x})$, to be minimized, is LP computable.

In Fig. 2.2, the values of the rate of return for a given portfolio under $T = 25$ scenarios are shown. The length of the vertical segment is the absolute value of the

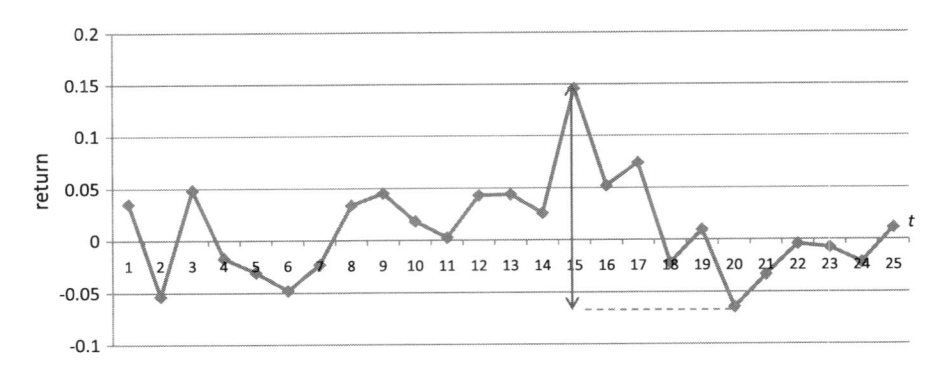

Fig. 2.2 The GMD measure

difference between the portfolio returns under scenarios 15 and 20, i.e. $d_{15,20} = |y_{15} - y_{20}|$.

The portfolio optimization model based on the GMD risk measure can be written as follows:

$$\min \sum_{t'=1}^{T} \sum_{t'' \neq t'} p_{t'} p_{t''} d_{t't''} \tag{2.12a}$$

$$d_{t't''} \geq y_{t'} - y_{t''} \qquad t', t'' = 1, \ldots, T;\ t'' \neq t' \tag{2.12b}$$

$$y_t = \sum_{j=1}^{n} r_{jt} x_j \qquad t = 1, \ldots, T \tag{2.12c}$$

$$\mu = \sum_{j=1}^{n} \mu_j x_j \tag{2.12d}$$

$$\mu \geq \mu_0 \tag{2.12e}$$

$$d_{t't''} \geq 0 \qquad t', t'' = 1, \ldots, T;\ t'' \neq t' \tag{2.12f}$$

$$\mathbf{x} \in Q. \tag{2.12g}$$

The model contains $T(T-1)$ non-negative variables $d_{t't''}$ and $T(T-1)$ inequalities to define them. The symmetry property $d_{t't''} = d_{t''t'}$ is here ignored. However, variables $d_{t't''}$ are associated with the singleton coefficient columns. Hence, while solving the dual instead of the original primal problem, the corresponding dual constraints take the form of simple upper bounds which are handled implicitly by the simplex method. In other words, the dual problem contains $T(T-1)$ variables but the number of constraints is then proportional to T. Such a dual approach may dramatically improve the required computational time in the case of large number of scenarios.

2.4 Basic LP Computable Safety Measures

Similarly to MAD, in the case when the rates of return are multivariate normally distributed, the proportionality relation $\Gamma(\mathbf{x}) = \frac{2}{\sqrt{\pi}}\sigma(\mathbf{x})$ between the Gini's mean difference and the standard deviation occurs. As a consequence, minimizing the GMD is equivalent to minimizing the standard deviation, which means, in this specific case, the equivalence of the associated optimization problems. Albeit, the GMD model does not require any specific type of return distribution.

2.4 Basic LP Computable Safety Measures

In the previous chapter and in the previous section of this chapter, we have seen some specific risk measures, the variance (Markowitz model), the mean absolute deviation (MAD), the Gini's mean difference (GMD). These measures capture, in different ways, the variability of the rate of return of the portfolio. Given a required expected return of the portfolio μ_0, the investor may wish to reduce the variability of the portfolio rate of return, that is to minimize any of these risk measures. We analyze here different ways to measure the quality of a portfolio and define some specific safety measures, to be maximized. We do not consider the variability of the portfolio rate of return, neither the deviations from its expected value. In fact, we ignore the expected rate of return and try instead to protect the investor from the worst scenarios.

An appealing safety measure is the worst realization of the portfolio rate of return. We aim at maximizing the worst realization of the portfolio rate of return. The *worst realization* is defined as

$$M(\mathbf{x}) = \min_{t=1,\ldots,T} y_t = \min_{t=1,\ldots,T} \sum_{j=1}^{n} r_{jt}x_j, \qquad (2.13)$$

and is LP computable. The portfolio optimization model with the worst realization as safety measure (the Minimax model) can be formulated as:

$$\max y \qquad (2.14a)$$

$$\sum_{j=1}^{n} r_{jt}x_j \geq y \qquad t = 1,\ldots,T \qquad (2.14b)$$

$$\mu = \sum_{j=1}^{n} \mu_j x_j \qquad (2.14c)$$

$$\mu \geq \mu_0 \qquad (2.14d)$$

$$\mathbf{x} \in Q. \qquad (2.14e)$$

The variable y is an artificial variable that in the optimum takes the value of the portfolio rate of return in the worst scenario. In Fig. 2.3, the rates of return for a given portfolio over 25 scenarios are drawn, and the worst realization of the portfolio rate of return is emphasized.

Suppose that, among the feasible portfolios of the Minimax model, there are the two portfolios shown in Table 2.2. Suppose that the required expected rate of return is $\mu_0 = 2\%$. Both portfolios $\mathbf{x}'$ and $\mathbf{x}''$ guarantee an expected rate of return not worse than 2%. Whereas portfolio $\mathbf{x}'$ has a larger expected rate of return, the model would prefer portfolio $\mathbf{x}''$ to portfolio $\mathbf{x}'$ because portfolio $\mathbf{x}''$ has the rate of return in the worst scenario, 2%, larger than the worst rate of return of portfolio $\mathbf{x}'$, 1.8 %. The maximization of the worst realization somehow pushes all the realizations toward larger – and thus better – values, but at the same time focuses on the worst scenario only.

A natural generalization of the measure $M(\mathbf{x})$ is the statistical concept of *quantile*. In general, for given $\beta \in [0, 1]$, the β-quantile of a random variable R is the value q such that

$$\mathbb{P}\{R < q\} \leq \beta \leq \mathbb{P}\{R \leq q\}.$$

For $\beta \in (0, 1)$, it is known that the set of such β-quantiles is a closed interval (see Embrechts et al. 1997). Given a value of β, in order to formalize the quantile

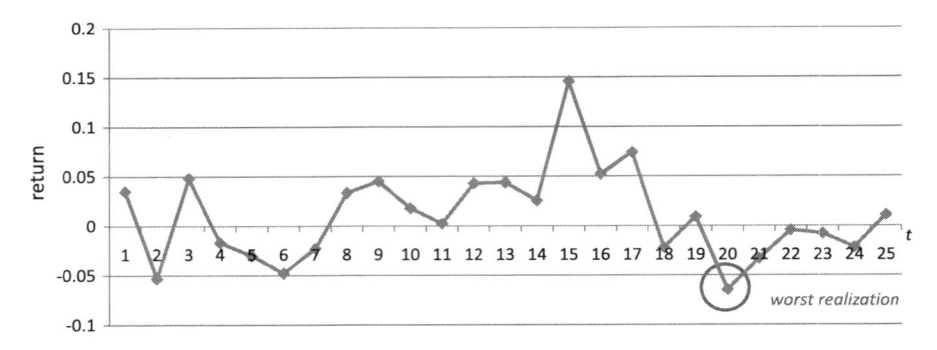

Fig. 2.3 The worst realization measure

Table 2.2 Optimal portfolio for the worst realization safety measure: An example

Scenario	Probability	Rates of return Portfolio $\mathbf{x}'$ (%)	Portfolio $\mathbf{x}''$ (%)
1	0.2	4.9	2.0
2	0.5	4.0	3.0
3	0.2	2.2	2.0
4	0.1	1.8	2.0
Mean	$\mu(\mathbf{x})$	3.6	2.5
Worst	$M(\mathbf{x})$	1.8	2.0

2.4 Basic LP Computable Safety Measures

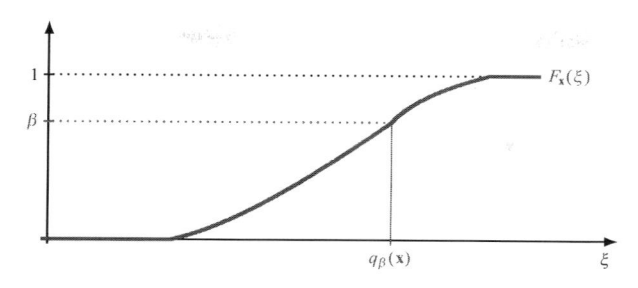

Fig. 2.4 VaR measure $q_\beta(\mathbf{x})$ and the cdf of portfolio returns

measures in the case of non-unique quantile, the left end of the entire interval of quantiles is taken. In our case, we denote by $q_\beta(\mathbf{x})$ the value of the β-quantile, that is the value of the rate of return defined as

$$q_\beta(\mathbf{x}) = \inf \{\eta : F_\mathbf{x}(\eta) \geq \beta\} \quad \text{for } 0 < \beta \leq 1, \tag{2.15}$$

where $F_\mathbf{x}(\cdot)$ is the cumulative distribution function defined in (2.1) (see Fig. 2.4).

In finance and banking literature, this quantile is usually called *Value-at-Risk* or simply *VaR* measure. Actually, for a given portfolio $\mathbf{x}$, its VaR depicts the worst (maximum) loss within a given confidence interval (see Jorion 2006). However, with a change of sign (losses as negative returns $-R_\mathbf{x}$), it is equivalent to the quantile $q_\beta(\mathbf{x})$.

Due to possible discontinuity of the cdf, the VaR measure is, generally, not an LP computable measure. The corresponding portfolio optimization model can be formulated as a MILP problem:

$$\max y \tag{2.16a}$$

$$\sum_{j=1}^{n} r_{jt} x_j \geq y - M z_t \qquad t = 1, \ldots, T \tag{2.16b}$$

$$\sum_{t=1}^{T} p_t z_t \leq \beta - \pi, \ z_t \in \{0, 1\} \qquad t = 1, \ldots, T \tag{2.16c}$$

$$\mu = \sum_{j=1}^{n} \mu_j x_j \tag{2.16d}$$

$$\mu \geq \mu_0 \tag{2.16e}$$

$$\mathbf{x} \in Q, \tag{2.16f}$$

where M is an arbitrary large constant (larger than any possible rate of return) while π is an arbitrary small positive constant ($\pi < p_t, t = 1, \ldots, T$). Note that, due to inequality (2.16b), binary variable z_t takes value 1 whenever variable y is greater than the portfolio return under scenario t ($y > y_t = \sum_{j=1}^{n} r_{jt}x_j$). Inequality (2.16c) guarantees that the probability of all scenarios such that $y > y_t$ is less than β. Therefore, the optimal value of the maximized variable y represents the optimal β-quantile value $q_\beta(\mathbf{x})$.

Recently, risk measures based on averaged quantiles have been introduced in different ways. The *tail mean* or *worst conditional expectation* $M_\beta(\mathbf{x})$, defined as the mean return of the portfolio taken over a given tolerance level (percentage) $0 < \beta \leq 1$ of the worst scenarios probability is a natural generalization of the measure $M(\mathbf{x})$. In finance literature, the tail mean quantity is usually called *Tail VaR*, *Average VaR* or *Conditional VaR* (where VaR reads after Value-at-Risk). Actually, the name CVaR is now the most commonly used and we adopt it.

For the simplest case of equally probable scenarios ($p_t = 1/T$) and proportional $\beta = k/T$, the CVaR measure $M_\beta(\mathbf{x})$ is defined as average of the k worst realizations

$$M_{\frac{k}{T}}(\mathbf{x}) = \frac{1}{k} \sum_{s=1}^{k} y_{t_s}, \qquad (2.17)$$

where $y_{t_1}, y_{t_2}, \ldots, y_{t_k}$ are the k worst realizations for the portfolio rate of return.

In Fig. 2.5, we show an example of a portfolio whose CVaR value has been computed for $k = 3$ and $T = 25$.

For any probability p_t and tolerance level β, due to the finite number of scenarios, the CVaR measure $M_\beta(\mathbf{x})$ is well defined by the following optimization

$$M_\beta(\mathbf{x}) = \min_{u_t}\{ \frac{1}{\beta} \sum_{t=1}^{T} y_t u_t : \sum_{t=1}^{T} u_t = \beta, \ 0 \leq u_t \leq p_t \ \ t = 1, \ldots, T\}, \qquad (2.18)$$

where at optimality u_t is the percentage of the t-th worst return in $M_\beta(\mathbf{x})$. More precisely, $u_t = 0$ for any scenario t not included in the worst scenarios, $u_t = p_t$

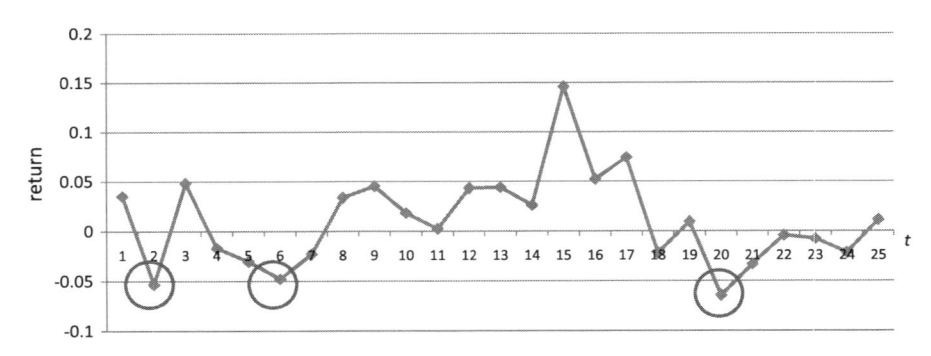

Fig. 2.5 The CVaR model $M_{\frac{k}{T}}(\mathbf{x})$

2.4 Basic LP Computable Safety Measures

Table 2.3 Optimal portfolios for the CVaR measure: An example

Scenario	Probability	Rates of return Portfolio x' (%)	Portfolio x'' (%)
1	0.2	4.9	2.0
2	0.5	4.0	3.0
3	0.2	2.2	2.0
4	0.1	1.8	2.0
Worst	$M(\mathbf{x})$	1.8	2.0
CVaR	$M_{0.05}(\mathbf{x})$	1.8	2.0
CVaR	$M_{0.1}(\mathbf{x})$	1.8	2.0
CVaR	$M_{0.2}(\mathbf{x})$	2.0	2.0
CVaR	$M_{0.3}(\mathbf{x})$	2.07	2.0
CVaR	$M_{0.5}(\mathbf{x})$	2.84	2.0
CVaR	$M_{0.8}(\mathbf{x})$	3.28	2.38
CVaR	$M_{1.0}(\mathbf{x})$	3.6	2.5
Mean	$\mu(\mathbf{x})$	3.6	2.5

for any scenario t totally included in the worst scenarios, and $0 < u_t < p_t$ for one scenario t only.

When parameter β approaches 0 and becomes smaller than or equal to the minimal scenario probability ($\beta \leq \min_t p_t$), the measure becomes the worst return $M(\mathbf{x}) = \lim_{\beta \to 0_+} M_\beta(\mathbf{x})$. On the other hand, for $\beta = 1$ the corresponding CVaR becomes the mean ($M_1(\mathbf{x}) = \mu(\mathbf{x})$).

Recall the case of two portfolios shown in Table 2.2. In Table 2.3, we show their CVaR values for various tolerance levels. For $\beta = 0.05$ and $\beta = 0.1$ the CVaR values are equal to the corresponding return in the worst scenario, $M(\mathbf{x}') = 1.8\%$ and $M(\mathbf{x}'') = 2\%$, respectively. For $\beta = 0.2$ one gets equal CVaR values $M_{0.2}(\mathbf{x}') = M_{0.2}(\mathbf{x}'') = 2\%$, while for $\beta = 0.3$ one has $M_{0.3}(\mathbf{x}') = 2.07\%$ greater than $M_{0.3}(\mathbf{x}'') = 2\%$. The difference becomes larger for tolerance levels $\beta = 0.5$ and $\beta = 0.8$. Obviously, for $\beta = 1$ one gets the corresponding means as CVaR values.

Problem (2.18) is a linear program for a given portfolio $\mathbf{x}$, while it becomes nonlinear when the y_t are variables in the portfolio optimization problem. It turns out that this difficulty can be overcome by taking advantage of the LP dual problem to (2.18) leading to an equivalent LP dual formulation of the CVaR model that allows one to implement the optimization problem with auxiliary linear inequalities. Indeed, introducing dual variable η corresponding to the equation $\sum_{t=1}^{T} u_t = \beta$ and variables d_t^- corresponding to upper bounds on u_t one gets the LP dual problem:

$$M_\beta(\mathbf{x}) = \max_{\eta, d_t^-} \{\eta - \frac{1}{\beta}\sum_{t=1}^{T} p_t d_t^- : d_t^- \geq \eta - y_t, \ d_t^- \geq 0 \ t = 1, \ldots, T\}. \quad (2.19)$$

Due to the duality theory, for any given vector y_t the measure $M_\beta(\mathbf{x})$ can be found as the optimal value of the LP problem (2.19). Thus, the CVaR is a safety measure that is LP computable. The portfolio optimization model can be formulated as follows:

$$\max \left(\eta - \frac{1}{\beta} \sum_{t=1}^{T} p_t d_t^-\right) \tag{2.20a}$$

$$d_t^- \geq \eta - y_t \qquad t = 1, \ldots, T \tag{2.20b}$$

$$y_t = \sum_{j=1}^{n} r_{jt} x_j \qquad t = 1, \ldots, T \tag{2.20c}$$

$$\mu = \sum_{j=1}^{n} \mu_j x_j \tag{2.20d}$$

$$\mu \geq \mu_0 \tag{2.20e}$$

$$d_t^- \geq 0 \qquad t = 1, \ldots, T \tag{2.20f}$$

$$\mathbf{x} \in Q, \tag{2.20g}$$

where η is an auxiliary (unbounded) variable that in the optimal solution will take the value of the β-quantile.

In the case of $\mathbb{P}\{R_\mathbf{x} \leq q_\beta(\mathbf{x})\} = \beta$, one gets $M_\beta(\mathbf{x}) = \mathbb{E}\{R_\mathbf{x} | R_\mathbf{x} \leq q_\beta(\mathbf{x})\}$. This represents the original concept of the CVaR measure. Although valid for many continuous distributions this formula, in general, cannot serve as a definition of the CVaR measure because a value ξ such that $\mathbb{P}\{R_\mathbf{x} \leq \xi\} = \beta$ may not exist. In general, $\mathbb{P}\{R_\mathbf{x} \leq q_\beta(\mathbf{x})\} = \beta' \geq \beta$ and $M_\beta(\mathbf{x}) \leq M_{\beta'}(\mathbf{x}) = \mathbb{E}\{R_\mathbf{x} | R_\mathbf{x} \leq q_\beta(\mathbf{x})\}$.

2.5 The Complete Set of Basic Linear Models

As shown in the previous sections several LP computable risk measures have been considered for portfolio optimization. Some of them were originally introduced rather as safety measures in our classification (e.g., CVaR measures). Nevertheless, all of them can be represented with positively homogeneous and shift independent risk measures ϱ of classical Markowitz type model. Simple as well as more complicated LP computable risk measures $\varrho(\mathbf{x})$ can be defined as

$$\varrho(\mathbf{x}) = \min\{\mathbf{a}^T \mathbf{v} : \quad \mathbf{A}\mathbf{v} = \mathbf{B}\mathbf{x}, \ \mathbf{v} \geq \mathbf{0}, \mathbf{x} \in Q\}, \tag{2.21}$$

where $\mathbf{v}$ is a vector of auxiliary variables while the portfolio vector $\mathbf{x}$, apart from original portfolio constraints $\mathbf{x} \in Q$, only defines a parametric right hand side vector $\mathbf{b} = \mathbf{B}\mathbf{x}$. Obviously, the corresponding safety measures are given by a similar LP formula

$$\mu(\mathbf{x}) - \varrho(\mathbf{x}) = \max\{\sum_{j=1}^{n} \mu_j x_j - \mathbf{a}^T \mathbf{v} : \quad \mathbf{A}\mathbf{v} = \mathbf{B}\mathbf{x}, \ \mathbf{v} \geq \mathbf{0}, \mathbf{x} \in Q\}. \tag{2.22}$$

2.5 The Complete Set of Basic Linear Models

For each model of type (2.21), the mean-risk bounding approach (1.10) leads to the LP problem

$$\min_{\mathbf{x},\mathbf{v}}\{\mathbf{a}^T\mathbf{v} : \mathbf{A}\mathbf{v} = \mathbf{B}\mathbf{x}, \mathbf{v} \geq \mathbf{0}, \sum_{j=1}^{n}\mu_j x_j \geq \mu_0, \mathbf{x} \in Q\}, \quad (2.23)$$

while the mean-safety bounding approach (1.12) applied to (2.22) results in

$$\max_{\mathbf{x},\mathbf{v}}\{\sum_{j=1}^{n}\mu_j x_j - \mathbf{a}^T\mathbf{v} : \mathbf{A}\mathbf{v} = \mathbf{B}\mathbf{x}, \mathbf{v} \geq \mathbf{0}, \sum_{j=1}^{n}\mu_j x_j \geq \mu_0, \mathbf{x} \in Q\}. \quad (2.24)$$

Similarly, the trade-off analysis approach (1.13) results in the LP model

$$\max_{\mathbf{x},\mathbf{v}}\{\sum_{j=1}^{n}\mu_j x_j - \lambda \mathbf{a}^T\mathbf{v} : \mathbf{A}\mathbf{v} = \mathbf{B}\mathbf{x}, \mathbf{v} \geq \mathbf{0}, \mathbf{x} \in Q\}. \quad (2.25)$$

2.5.1 Risk Measures from Safety Measures

Recall that, for a discrete random variable represented by its realizations y_t, the *worst realization* $M(\mathbf{x}) = \min_{t=1,\ldots,T}\{y_t\}$ is an appealing LP computable safety measure (see (2.13)). The corresponding (dispersion) risk measure $\Delta(\mathbf{x}) = \mu(\mathbf{x}) - M(\mathbf{x})$, the *maximum (downside) semideviation*, is LP computable as

$$\Delta(\mathbf{x}) = \min\{v : v \geq \sum_{j=1}^{n}(\mu_j - r_{jt})x_j, t = 1,\ldots,T\}. \quad (2.26)$$

The portfolio optimization model with the maximum semideviation as risk measure can be formulated as:

$$\min v \quad (2.27a)$$

$$\mu - \sum_{j=1}^{n} r_{jt} x_j \leq v \qquad t = 1,\ldots,T \quad (2.27b)$$

$$\mu = \sum_{j=1}^{n}\mu_j x_j \quad (2.27c)$$

$$\mu \geq \mu_0 \quad (2.27d)$$

$$\mathbf{x} \in Q. \quad (2.27e)$$

The variable v is an auxiliary variable that in the optimum will take the value of the maximum downside deviation of the portfolio rate of return from the mean return.

Similarly, the CVaR measure is a safety measure. The corresponding risk measure $\Delta_\beta(\mathbf{x}) = \mu(\mathbf{x}) - M_\beta(\mathbf{x})$ is called the (worst) *conditional semideviation* or *conditional drawdown* measure. For a discrete random variable represented by its realizations, due to (2.19), the conditional semideviations may be computed as the corresponding differences from the mean:

$$\Delta_\beta(\mathbf{x}) = \min\{\sum_{j=1}^{n} \mu_j x_j - \eta + \frac{1}{\beta} \sum_{t=1}^{T} d_t^- p_t : d_t^- \geq \eta - y_t, \ d_t^- \geq 0, \ t = 1, \ldots, T\}, \tag{2.28}$$

or, equivalently, setting $d_t^+ = d_t^- - \eta + y_t$, as:

$$\Delta_\beta(\mathbf{x}) = \min\{\sum_{t=1}^{T}(d_t^+ + \frac{1-\beta}{\beta} d_t^-) p_t : d_t^- - d_t^+ = \eta - y_t, \ d_t^-, d_t^+ \geq 0, \ t = 1, \ldots, T\}, \tag{2.29}$$

where η is an auxiliary (unbounded) variable that in the optimal solution will take the value of the β-quantile $q_\beta(\mathbf{x})$.

Thus, the conditional semideviation is an LP computable risk measure and the corresponding portfolio optimization model can be formulated as follows:

$$\min \ \sum_{t=1}^{T}(d_t^+ + \frac{1-\beta}{\beta} d_t^-) p_t \tag{2.30a}$$

$$d_t^- - d_t^+ = \eta - y_t, \ d_t^-, d_t^+ \geq 0 \qquad t = 1, \ldots, T \tag{2.30b}$$

$$y_t = \sum_{j=1}^{n} r_{jt} x_j \qquad t = 1, \ldots, T \tag{2.30c}$$

$$\mu = \sum_{j=1}^{n} \mu_j x_j \tag{2.30d}$$

$$\mu \geq \mu_0 \tag{2.30e}$$

$$\mathbf{x} \in Q. \tag{2.30f}$$

Note that for $\beta = 0.5$ one has $(1 - \beta)/\beta = 1$. Hence, $\Delta_{0.5}(\mathbf{x})$ represents the mean absolute deviation from the median $q_{0.5}(\mathbf{x})$. The LP problem for computing

2.5 The Complete Set of Basic Linear Models

this measure can be expressed in the form:

$$\Delta_{0.5}(\mathbf{x}) = \min\{\sum_{t=1}^{T} d_t p_t : d_t \geq \eta - y_t, \ d_t \geq y_t - \eta, \ d_t \geq 0 \quad t = 1, \ldots, T\}.$$

One may notice that the above model differs from the classical MAD formulation (2.8) only due to replacement of $\mu(\mathbf{x})$ with (unrestricted) variable η.

2.5.2 Safety Measures from Risk Measures

Symmetrically, a safety measure can be obtained from a positively homogeneous and shift independent (deviation type) risk measure. For the Semi-MAD (2.9) the corresponding safety measure can be expressed as

$$\mu(\mathbf{x}) - \bar{\delta}(\mathbf{x}) = \mathbb{E}\{\mu(\mathbf{x}) - \max\{\mu(\mathbf{x}) - R_\mathbf{x}, 0\}\} = \mathbb{E}\{\min\{R_\mathbf{x}, \mu(\mathbf{x})\}\}, \qquad (2.31)$$

thus representing the *mean downside underachievement*. The corresponding portfolio optimization problem can be written as follows:

$$\max \sum_{t=1}^{T} p_t v_t \qquad (2.32a)$$

$$v_t \leq \sum_{j=1}^{n} r_{jt} x_j \qquad t = 1, \ldots, T \qquad (2.32b)$$

$$v_t \leq \mu \qquad t = 1, \ldots, T \qquad (2.32c)$$

$$\mu = \sum_{j=1}^{n} \mu_j x_j \qquad (2.32d)$$

$$\mu \geq \mu_0 \qquad (2.32e)$$

$$\mathbf{x} \in Q. \qquad (2.32f)$$

The Gini's mean difference (2.11) has the corresponding safety measure defined as

$$\mu(\mathbf{x}) - \Gamma(\mathbf{x}) = \sum_{t'=1}^{T} \sum_{t''=1}^{T} \min\{y_{t'}, y_{t''}\} p_{t'} p_{t''}, \qquad (2.33)$$

where the latter expression is obtained through algebraic calculations. Hence, (2.33) is the expectation of the minimum of two independent identically distributed random variables, thus representing the *mean worse return*.

This leads to the following LP formula

$$\mu(\mathbf{x}) - \Gamma(\mathbf{x}) = \max\{\sum_{t'=1}^{T}\sum_{t''=1}^{T} v_{t't''} p_{t'} p_{t''} : \\ v_{t't''} \leq \sum_{j=1}^{n} r_{jt'} x_j, \ v_{t't''} \leq \sum_{j=1}^{n} r_{jt''} x_j, t', t'' = 1, \ldots, T\}. \quad (2.34)$$

The portfolio optimization model can be written as follows:

$$\max \sum_{t'=1}^{T}\sum_{t''=1}^{T} p_{t'} p_{t''} v_{t't''} \quad (2.35a)$$

$$v_{t't''} \leq y_{t'} \qquad t', t'' = 1, \ldots, T \quad (2.35b)$$

$$v_{t't''} \leq y_{t''} \qquad t', t'' = 1, \ldots, T \quad (2.35c)$$

$$y_t = \sum_{j=1}^{n} r_{jt} x_j \qquad t = 1, \ldots, T \quad (2.35d)$$

$$\mu = \sum_{j=1}^{n} \mu_j x_j \quad (2.35e)$$

$$\mu \geq \mu_0 \quad (2.35f)$$

$$\mathbf{x} \in Q. \quad (2.35g)$$

2.5.3 Ratio Measures from Risk Measures

As mentioned in Chap. 1, an alternative approach to the bicriteria mean-risk approach to portfolio selection is based on maximizing the ratio $(\mu(\mathbf{x}) - r_0)/\varrho(\mathbf{x})$. The corresponding ratio optimization problem (1.14) can be converted into an LP form by the following transformation: introduce an auxiliary variable $z = 1/\varrho(\mathbf{x})$, then replace the original variables $\mathbf{x}$ and $\mathbf{v}$ with $\tilde{\mathbf{x}} = z\mathbf{x}$ and $\tilde{\mathbf{v}} = z\mathbf{v}$, respectively, getting the linear criterion and an LP feasible set. For risk measure $\varrho(\mathbf{x})$ defined by (2.21) one gets the following LP formulation of the corresponding ratio model

$$\max_{\tilde{\mathbf{x}}, \tilde{\mathbf{v}}, z}\{\sum_{j=1}^{n} \mu_j \tilde{x}_j - r_0 z \ : \ \mathbf{c}^T \tilde{\mathbf{v}} = z, \ \mathbf{A}\tilde{\mathbf{v}} = \mathbf{b}\tilde{\mathbf{x}}, \ \tilde{\mathbf{v}} \geq \mathbf{0}, \\ \sum_{j=1}^{n} \tilde{x}_j = z, \ \tilde{x}_j \geq 0, j = 1, \ldots, n\}, \quad (2.36)$$

2.5 The Complete Set of Basic Linear Models

where the second line constraints correspond to the simplest definition of set $Q = \{\mathbf{x} : \sum_{j=1}^{n} x_j = 1, \ x_j \geq 0, j = 1, \ldots, n\}$ and can be accordingly formulated for any other LP set. Once the transformed problem is solved, the values of the portfolio variables x_j can be found by dividing $\tilde{x}_j$ by the optimal value of z. Thus, the LP computable portfolio optimization models, we consider, remain within LP environment even in the case of ratio criterion used to define the tangency portfolio.

For the Semi-MAD model (2.10) with risk measure $\varrho(\mathbf{x}) = \bar{\delta}(\mathbf{x})$, the ratio optimization model can be written as

$$\max \left\{ \frac{\mu - r_0}{\sum_{t=1}^{T} p_t d_t} : (2.10\text{b})\text{–}(2.10\text{g}) \right\}.$$

Introducing variables $z = 1/\sum_{t=1}^{T} p_t d_t$ and $\tilde{v} = z\mu$ we get the linear criterion $\tilde{v} - r_0 z$. Further, we multiply all the constraints by z and make the substitutions: $\tilde{d}_t = z d_t$, $\tilde{y}_t = z y_t$, for $t = 1, \ldots, T$, as well as $\tilde{x}_j = z x_j$, for $j = 1, \ldots, n$. Finally, we get the following LP formulation:

$$\max \tilde{v} - r_0 z \tag{2.37a}$$

$$\sum_{t=1}^{T} p_t \tilde{d}_t = 1 \tag{2.37b}$$

$$\tilde{d}_t \geq \tilde{v} - \tilde{y}_t, \ \tilde{d}_t \geq 0 \qquad t = 1, \ldots, T \tag{2.37c}$$

$$\tilde{y}_t = \sum_{j=1}^{n} r_{jt} \tilde{x}_j \qquad t = 1, \ldots, T \tag{2.37d}$$

$$\tilde{v} = \sum_{j=1}^{n} \mu_j \tilde{x}_j \tag{2.37e}$$

$$\sum_{j=1}^{n} \tilde{x}_j = z, \quad \tilde{x}_j \geq 0 \qquad j = 1, \ldots, n, \tag{2.37f}$$

where the last constraints correspond to the simplest definition of set Q.

Clear identification of dispersion type risk measures for all the LP computable safety measures allows us to define tangency portfolio optimization for all the models.

For the CVaR model with conditional semideviation as risk measure $\varrho(\mathbf{x}) = \Delta_\beta(\mathbf{x})$ (2.30) the ratio optimization model can be written as

$$\max \left\{ \frac{\mu - r_0}{\sum_{t=1}^{T} (d_t^+ + \frac{1-\beta}{\beta} d_t^-) p_t} : (2.30\text{b})\text{–}(2.30\text{f}) \right\}.$$

Introducing variables $z = 1/\sum_{t=1}^{T}(d_t^+ + \frac{1-\beta}{\beta}d_t^-)p_t$ and $\tilde{v} = z\mu$ we get the linear criterion $\tilde{v} - r_0 z$. Further, we multiply all the constraints by z and make the substitutions: $\tilde{d}_t^+ = zd_t^+$, $\tilde{d}_t^- = zd_t^-$, $\tilde{y}_t = zy_t$ for $t = 1,\ldots,T$, as well as $\tilde{x}_j = zx_j$, for $j = 1,\ldots,n$. Then, we get the following LP formulation:

$$\max \tilde{v} - r_0 z \tag{2.38a}$$

$$\sum_{t=1}^{T}(\tilde{d}_t^+ + \frac{1-\beta}{\beta}\tilde{d}_t^-)p_t = 1 \tag{2.38b}$$

$$\tilde{d}_t^- - \tilde{d}_t^+ = \eta - \tilde{y}_t, \quad \tilde{d}_t^-, \tilde{d}_t^+ \geq 0 \qquad t = 1,\ldots,T \tag{2.38c}$$

$$\tilde{y}_t = \sum_{j=1}^{n} r_{jt}\tilde{x}_j \qquad t = 1,\ldots,T \tag{2.38d}$$

$$\tilde{v} = \sum_{j=1}^{n} \mu_j \tilde{x}_j \tag{2.38e}$$

$$\sum_{j=1}^{n} \tilde{x}_j = z, \quad \tilde{x}_j \geq 0 \qquad j = 1,\ldots,n. \tag{2.38f}$$

2.6 Advanced LP Computable Measures

The LP computable risk measures may be further extended to enhance the risk aversion modeling capabilities. First of all, the measures may be combined in a weighted sum which allows the generation of various mixed measures. Consider a set of, say m, risk measures $\varrho_k(\mathbf{x})$ and their linear combination with weights w_k:

$$\varrho_{\mathbf{w}}^{(m)}(\mathbf{x}) = \sum_{k=1}^{m} w_k \varrho_k(\mathbf{x}), \qquad \sum_{k=1}^{m} w_k \leq 1, \quad w_k \geq 0 \quad \text{for } k = 1,\ldots,m. \tag{2.39}$$

Note that

$$\mu(\mathbf{x}) - \varrho_{\mathbf{w}}^{(m)}(\mathbf{x}) = w_0 \mu(\mathbf{x}) + \sum_{k=1}^{m} w_k(\mu(\mathbf{x}) - \varrho_k(\mathbf{x})),$$

where $w_0 = 1 - \sum_{k=1}^{m} w_k \geq 0$.

2.6 Advanced LP Computable Measures

In particular, one may build a multiple CVaR measure by considering, say m, tolerance levels $0 < \beta_1 < \beta_2 < \ldots < \beta_m \leq 1$ and using the weighted sum of the conditional semideviations $\Delta_{\beta_k}(\mathbf{x})$ as a new risk measure

$$\Delta_{\mathbf{w}}^{(m)}(\mathbf{x}) = \sum_{k=1}^{m} w_k \Delta_{\beta_k}(\mathbf{x}), \quad \sum_{k=1}^{m} w_k = 1, \quad w_k > 0 \quad \text{for } k = 1, \ldots, m, \quad (2.40)$$

with the corresponding safety measure

$$M_{\mathbf{w}}^{(m)}(\mathbf{x}) = \mu(\mathbf{x}) - \Delta_{\mathbf{w}}^{(m)}(\mathbf{x}) = \sum_{k=1}^{m} w_k M_{\beta_k}(\mathbf{x}). \quad (2.41)$$

The resulting Weighted CVaR (WCVaR) models use multiple CVaR measures, thus allowing for more detailed risk aversion modeling. The WCVaR risk measure is obviously LP computable as

$$\begin{aligned}
M_{\mathbf{w}}^{(m)}(\mathbf{x}) = \max \{ &\sum_{k=1}^{m} w_k (\eta_k - \frac{1}{\beta_k} \sum_{t=1}^{T} d_{kt}^{-} p_t) : d_{kt}^{-} \geq 0, \\
&d_{kt}^{-} \geq \eta_k - \sum_{j=1}^{n} r_{jt} x_j, t = 1, \ldots, T; k = 1, \ldots, m \}.
\end{aligned} \quad (2.42)$$

The corresponding portfolio optimization model can be formulated as follows:

$$\max \sum_{k=1}^{m} w_k (\eta_k - \frac{1}{\beta_k} \sum_{t=1}^{T} d_{kt}^{-} p_t) \quad (2.43a)$$

$$d_{kt}^{-} \geq \eta_k - y_t, \ d_{kt}^{-} \geq 0 \qquad t = 1, \ldots, T; k = 1, \ldots, m \quad (2.43b)$$

$$y_t = \sum_{j=1}^{n} r_{jt} x_j \qquad t = 1, \ldots, T \quad (2.43c)$$

$$\mu = \sum_{j=1}^{n} \mu_j x_j \quad (2.43d)$$

$$\mu \geq \mu_0 \quad (2.43e)$$

$$\mathbf{x} \in Q. \quad (2.43f)$$

For appropriately defined weights the WCVaR measures may be considered some approximations to the Gini's mean difference with the advantage of being computationally much simpler than the GMD model itself.

The risk measures introduced in the previous section are quite different in modeling the downside risk aversion. Definitely, the strongest in this respect is the maximum semideviation $\Delta(\mathbf{x})$ while the conditional semideviation $\Delta_\beta(\mathbf{x})$ (CVaR model) allows us to extend the approach to a specified β quantile of the worst returns which results in a continuum of models evolving from the strongest downside risk aversion (β close to 0) to the complete risk neutrality ($\beta = 1$). The mean (downside) semideviation from the mean, used in the MAD model, is formally a downside risk measure. However, due to the symmetry of mean semideviations from the mean it is equally appropriate to interpret it as a measure of the upside risk. Similarly, the Gini's mean difference, although related to all the conditional maximum semideviations, is a symmetric risk measure (in the sense that for $R_\mathbf{x}$ and $-R_\mathbf{x}$ it has exactly the same value). For better modeling of the risk averse preferences one may enhance the below-mean downside risk aversion in various measures. The below-mean risk downside aversion is a concept of risk aversion assuming that the variability of returns above the mean should not be penalized since the investors are concerned about the under-performance rather than the over-performance of a portfolio. This can be implemented by focusing on the distribution of *downside underachievements* $\min\{R_\mathbf{x}, \mu(\mathbf{x})\}$ instead of the original distribution of returns $R_\mathbf{x}$.

Applying the mean semideviation (2.9) to the distribution of downside underachievements $\min\{R_\mathbf{x}, \mu(\mathbf{x})\}$ one gets

$$\bar{\delta}_2(\mathbf{x}) = \mathbb{E}\{\max\{\mathbb{E}\{\min\{R_\mathbf{x}, \mu(\mathbf{x})\}\} - R_\mathbf{x}, 0\}\} = \mathbb{E}\{\max\{\mu(\mathbf{x}) - \bar{\delta}(\mathbf{x}) - R_\mathbf{x}, 0\}\}.$$

This allows us to define the enhanced risk measure for the original distribution of returns $R_\mathbf{x}$ as $\bar{\delta}^{(2)}(\mathbf{x}) = \bar{\delta}(\mathbf{x}) + \bar{\delta}_2(\mathbf{x})$ with the corresponding safety measure $\mu(\mathbf{x}) - \bar{\delta}^{(2)}(\mathbf{x}) = \mu(\mathbf{x}) - \bar{\delta}(\mathbf{x}) - \bar{\delta}_2(\mathbf{x})$. The above approach can be repeated recursively, resulting in m (defined recursively) distribution dependent targets $\mu_1(\mathbf{x}) = \mu(\mathbf{x})$, $\mu_k(\mathbf{x}) = \mathbb{E}\{\min\{R_\mathbf{x}, \mu_{k-1}(\mathbf{x})\}\}$ for $k = 2,\ldots,m$, and the corresponding mean semideviations $\bar{\delta}_1(\mathbf{x}) = \bar{\delta}(\mathbf{x})$, $\bar{\delta}_k(\mathbf{x}) = \mathbb{E}\{\max\{\mu_k(\mathbf{x}) - R_\mathbf{x}, 0\}\}$ for $k = 1,\ldots,m$. The measure

$$\bar{\delta}_\mathbf{w}^{(m)}(\mathbf{x}) = \sum_{k=1}^{m} w_k \bar{\delta}_k(\mathbf{x}) \qquad 1 = w_1 \geq w_2 \geq \ldots \geq w_m \geq 0 \qquad (2.44)$$

gives rise to the so-called *m*-MAD model. Actually, the measure can be interpreted as a single mean semideviation (from the mean) applied with a penalty function: $\bar{\delta}_\mathbf{w}^{(m)}(\mathbf{x}) = \mathbb{E}\{u(\max\{\mu(\mathbf{x}) - R_\mathbf{x}, 0\})\}$, where u is an increasing and convex piecewise linear penalty function with breakpoints $b_k = \mu(\mathbf{x}) - \mu_k(\mathbf{x})$ and slopes $s_k = w_1 + \ldots + w_k$, $k = 1,\ldots,m$. Therefore, the measure $\bar{\delta}_\mathbf{w}^{(m)}(\mathbf{x})$ is referred to as the *mean penalized semideviation* and is obviously LP computable leading to the following LP form of

2.6 Advanced LP Computable Measures

the m-MAD portfolio optimization model:

$$\min \sum_{k=1}^{m} w_k v_k \tag{2.45a}$$

$$v_k - \sum_{t=1}^{T} p_t d_{kt} = 0 \qquad k = 1, \ldots, m \tag{2.45b}$$

$$d_{kt} \geq 0, \ d_{kt} \geq \mu - y_t - \sum_{i=1}^{k-1} v_i \qquad t = 1, \ldots, T; \ k = 1, \ldots, m \tag{2.45c}$$

$$y_t = \sum_{j=1}^{n} r_{jt} x_j \qquad t = 1, \ldots, T \tag{2.45d}$$

$$\mu = \sum_{j=1}^{n} \mu_j x_j \tag{2.45e}$$

$$\mu \geq \mu_0 \tag{2.45f}$$

$$\mathbf{x} \in Q. \tag{2.45g}$$

The Gini's mean difference is a symmetric measure, thus equally treating both under and over-achievements. The enhancement technique allows us to define the downside Gini's mean difference by applying the Gini's mean difference to the distribution of downside underachievements $\min\{R_\mathbf{x}, \mu(\mathbf{x})\}$

$$\Gamma_2(\mathbf{x}) = \sum_{t=1}^{T} \min\{y_t, \mu(\mathbf{x})\} p_t - \sum_{t'=1}^{T} \sum_{t''=1}^{T} \min\{\min\{y_{t'}, \mu(\mathbf{x})\}, \min\{y_{t''}, \mu(\mathbf{x})\}\} p_{t'} p_{t''}.$$

Hence, we get the *downside Gini's mean difference* defined as the enhanced risk measure:

$$\Gamma^d(\mathbf{x}) = \Gamma_2(\mathbf{x}) + \bar{\delta}(\mathbf{x}) = \mu(\mathbf{x}) - \sum_{t'=1}^{T} \sum_{t''=1}^{T} \min\{y_{t'}, y_{t''}, \mu(\mathbf{x})\} p_{t'} p_{t''}. \tag{2.46}$$

The downside Gini's safety measure takes the form:

$$\mu(\mathbf{x}) - \Gamma^d(\mathbf{x}) = \sum_{t'=1}^{T} \sum_{t''=1}^{T} \min\{y_{t'}, y_{t''}, \mu(\mathbf{x})\} p_{t'} p_{t''}, \tag{2.47}$$

which is obviously LP computable. The portfolio optimization model based on the downside Gini's safety measure can be written as follows:

$$\max \sum_{t'=1}^{T} \sum_{t''=1}^{T} p_{t'} p_{t''} v_{t't''} \tag{2.48a}$$

$$v_{t't''} \leq y_{t'} \qquad t', t'' = 1, \ldots, T \tag{2.48b}$$

$$v_{t't''} \leq y_{t''} \qquad t', t'' = 1, \ldots, T \tag{2.48c}$$

$$v_{t't''} \leq \mu \qquad t', t'' = 1, \ldots, T \tag{2.48d}$$

$$y_t = \sum_{j=1}^{n} r_{jt} x_j \qquad t = 1, \ldots, T \tag{2.48e}$$

$$\mu = \sum_{j=1}^{n} \mu_j x_j \tag{2.48f}$$

$$\mu \geq \mu_0 \tag{2.48g}$$

$$\mathbf{x} \in Q. \tag{2.48h}$$

The notion of risk may be related to a possible failure of achieving some targets instead of the mean. It was formalized by the concept of below-target risk measures or shortfall criteria. The simplest shortfall criterion for a specific target value τ is the mean below-target deviation (first Lower Partial Moment, LPM)

$$\bar{\delta}_\tau(\mathbf{x}) = \mathbb{E}\{\max\{\tau - R_\mathbf{x}, 0\}\}. \tag{2.49}$$

The mean below-target deviation is LP computable for returns represented by their realizations and the corresponding portfolio optimization model can be written as follows:

$$\min \sum_{t=1}^{T} p_t d_t \tag{2.50a}$$

$$d_t \geq \tau - y_t \qquad t = 1, \ldots, T \tag{2.50b}$$

$$y_t = \sum_{j=1}^{n} r_{jt} x_j \qquad t = 1, \ldots, T \tag{2.50c}$$

$$\mu = \sum_{j=1}^{n} \mu_j x_j \tag{2.50d}$$

$$\mu \geq \mu_0 \tag{2.50e}$$

$$d_t \geq 0 \qquad t = 1, \ldots, T \tag{2.50f}$$

$$\mathbf{x} \in Q. \tag{2.50g}$$

2.6 Advanced LP Computable Measures

The mean below-target deviation from a specific target (2.49) represents only a single criterion. One may consider several, say m, targets $\tau_1 > \tau_2 > \ldots > \tau_m$ and use weighted sum of the shortfall criteria as a risk measure:

$$\sum_{k=1}^{m} w_k \bar{\delta}_{\tau_k}(\mathbf{x}) = \sum_{k=1}^{m} w_k \mathbb{E}\{\max\{\tau_k - R_\mathbf{x}, 0\}\}, \quad (2.51)$$

where w_k (for $k = 1, \ldots, m$) are positive weights which maintain the measure LP computable (when minimized). Actually, the measure can be interpreted as a single mean below-target deviation applied with a penalty function: $\mathbb{E}\{u(\max\{\tau_1 - R_\mathbf{x}, 0\})\}$, where u is increasing and convex piece-wise linear penalty function with breakpoints $b_k = \tau_1 - \tau_k$ and slopes $\bar{s}_k = w_1 + \ldots + w_k$, $k = 1, \ldots, m$.

The below-target deviations are very useful in investment situations with clearly defined minimum acceptable returns (e.g. bankruptcy level). Otherwise, appropriate selection of the target value might be a difficult task. However, for portfolio optimization they are rather rarely applied. Recently, the so-called Omega ratio measure defined, for a given target, as the ratio of the mean over-target deviation by the mean below-target deviation was introduced:

$$\Omega_\tau(\mathbf{x}) = \frac{\mathbb{E}\{\max\{R_\mathbf{x} - \tau, 0\}\}}{\mathbb{E}\{\max\{\tau - R_\mathbf{x}, 0\}\}} = \frac{\int_\tau^\infty (1 - F_\mathbf{x}(\xi))\, d\xi}{\int_{-\infty}^\tau F_\mathbf{x}(\xi)\, d\xi}. \quad (2.52)$$

Since $\tau - \mathbb{E}\{\max\{\tau - R_\mathbf{x}, 0\}\} = \mu(\mathbf{x}) - \mathbb{E}\{\max\{R_\mathbf{x} - \tau, 0\}\}$, one gets

$$\Omega_\tau(\mathbf{x}) = \frac{\bar{\delta}_\tau(\mathbf{x}) - (\tau - \mu(\mathbf{x}))}{\bar{\delta}_\tau(\mathbf{x})} = 1 + \frac{\mu(\mathbf{x}) - \tau}{\bar{\delta}_\tau(\mathbf{x})}.$$

Thus, the portfolio optimization model based on the Omega ratio maximization is equivalent to the standard ratio (tangent portfolio) model (1.14) for the $\bar{\delta}_\tau(\mathbf{x})$ measure with target τ replacing the risk-free rate of return:

$$\max \left\{ \frac{\mu - \tau}{\sum_{t=1}^T p_t d_t} : (2.50\text{b})\text{–}(2.50\text{g}) \right\}.$$

Similarly to the MAD ratio model it is easily transformed to an LP form. Introducing variables $z = 1/\sum_{t=1}^T p_t d_t$ and $\tilde{v} = z\mu$ we get the linear criterion $\tilde{v} - \tau z$. Further, we multiply all the constraints by z and make the substitutions: $\tilde{d}_t = zd_t$, $\tilde{y}_t = zy_t$ for

$t = 1, \ldots, T$, as well as $\tilde{x}_j = zx_j$, for $j = 1, \ldots, n$. Finally, we get the following LP formulation:

$$\max \tilde{v} - \tau z \tag{2.53a}$$

$$\sum_{t=1}^{T} p_t \tilde{d}_t = 1 \tag{2.53b}$$

$$\tilde{d}_t + \tilde{y}_t \geq \tau z, \ \tilde{d}_t \geq 0 \qquad t = 1, \ldots, T \tag{2.53c}$$

$$\sum_{j=1}^{n} \mu_j \tilde{x}_j = \tilde{v} \tag{2.53d}$$

$$\sum_{j=1}^{n} r_{jt} \tilde{x}_j = \tilde{y}_t \qquad t = 1, \ldots, T \tag{2.53e}$$

$$\sum_{j=1}^{n} \tilde{x}_j = z, \quad \tilde{x}_j \geq 0 \qquad j = 1, \ldots, n, \tag{2.53f}$$

where the last constraints correspond to the set Q definition.

2.7 Notes and References

Initial attempts to have portfolio optimization models depended on the piecewise linear approximation of the variance (see Sharpe 1971a; Stone 1973). Later, several LP computable risk measures were introduced. Yitzhaki (1982) proposed the LP solvable portfolio optimization mean-risk model using Gini's mean (absolute) difference as the risk measure (the GMD model). The mean absolute deviation was very early considered in portfolio analysis by Sharpe (1971b). The complete LP solvable portfolio optimization model based on this risk measure (the MAD model) was presented and analyzed by Konno and Yamazaki (1991). The MAD model was extensively tested on various stock markets (see Konno and Yamazaki 1991; Mansini et al. 2003a; Xidonas et al. 2010) including its application to portfolios of mortgage-backed securities by Zenios and Kang (1993) where the distribution of rates of return is known to be non-symmetric. The MAD model usually, similarly to the Markowitz one, generated the portfolios with the largest returns but also entailing the largest risk of underachievement. This model has generated interest in LP portfolio optimization resulting in many new developments. Young (1998) analyzed the LP solvable portfolio optimization model based on risk defined by the worst case scenario (Minimax model), while Ogryczak (2000) introduced the multiple criteria LP model covering all the above as special aggregation techniques.

The Semi-MAD was independently presented by Feinstein and Thapa (1993) and Speranza (1993). The m-MAD model were introduced by Michalowski and

2.7 Notes and References

Ogryczak (2001), while Krzemienowski and Ogryczak (2005) introduced the downside Gini's mean difference. The mean absolute deviation from the median was suggested as risk measure by Sharpe (1971a).

The quantile risk measures were introduced in different ways by many authors (see Artzner et al. 1999; Embrechts et al. 1997; Ogryczak 1999; Rockafellar and Uryasev 2000). The tail mean or worst conditional expectation, defined as the mean return of the portfolio taken over a given percentage of the worst scenarios is a natural generalization of the measure due to Young (1998). In financial literature, the tail mean quantity is usually called tail VaR, average VaR or Conditional VaR (CVaR) (see Pflug 2000). Actually, the name CVaR, after Rockafellar and Uryasev (2000), is now the most commonly used. The measure was studied in several applications (see Andersson et al. 2001; Krokhmal et al. 2002; Roman et al. 2007; Topaloglou et al. 2002), and expanded in various forms (see Acerbi 2002; Krzemienowski 2009; Mansini et al. 2007; Zhu and Fukushima 2009).

Formal classification into risk and safety measures and their complementary pairs was introduced in Mansini et al. (2003a). The maximum semideviation measure was introduced in Ogryczak (2000). The (deviation) risk measure corresponding to the CVaR was considered as the (worst) conditional semideviation (Ogryczak and Ruszczyński 2002a) or conditional drawdown measure (Chekhlov et al. 2005). General deviation risk measures were analyzed by Rockafellar et al. (2006). Linear formulations of the ratio optimization models for all the basic LP computable risk measures was introduced in Mansini et al. (2003b).

The notion of risk related to a possible failure of achieving some targets was introduced by Roy (1952) as the so-called safety-first strategy and later led to the concept of below-target risk measures (see Fishburn 1977; Nawrocki 1992) or shortfall criteria.

The Omega measure was introduced by Shadwick and Keating (2002), while the first LP portfolio optimization model with this measure was shown by Mausser et al. (2006).

Chapter 3
Portfolio Optimization with Transaction Costs

3.1 Introduction

Investors, individuals and financial institutions, tend to invest money in a relatively small number of assets. In general this is due to the management costs, and to the costs that are incurred when buying and selling an asset. In financial markets, the latter costs are commonly called *transaction costs* and include commissions and other costs charged by a broker or a financial institution that plays an intermediary role between the investor and the seller of the asset. Broadly speaking, the transaction costs are the payments that brokers and financial institutions receive for performing transactions (buying or selling assets). Transaction costs are important to investors because they are one of the key determinants of their net returns. Transaction costs diminish the net returns and reduce the amount of capital available for future investments.

Transaction costs may have different structures. To express the transaction costs in linear form a specific use of variables and constraints is requested. A model may account for transaction costs in different ways. For example, transaction costs may reduce the final net portfolio return and/or diminish the capital available for the actual investment. When a complete portfolio optimization model is defined, some of the constraints on the definition of the transaction costs may be relaxed without affecting the correctness of the model as the optimization 'pushes' the transaction costs to the minimum value allowed by the constraints.

We will first present the most common structures of transaction costs and see how to model them. Then we will discuss the different ways to account for transaction costs in a portfolio optimization model. Finally, we will present, as an example, a complete portfolio optimization model using CVaR as objective function.

In this chapter, we measure the quantity of an asset in a portfolio in terms of amount of capital, and use variables $X_j, j = 1, \ldots, n$. We define *available capital* the total amount of money that is available to an investor, both for the investment in assets and transaction costs. The *invested capital* is the capital strictly used for the investment and that yields a return. Frequently, in portfolio optimization models the invested capital coincides with the available capital, and thus is a constant. We indicate the available capital with $\bar{C}$.

Being a constant, capital $\bar{C}$ can be normalized to 1 and the amount invested in each asset can be expressed as share by the linear transformation $x_j = X_j/\bar{C}$. We have already clarified that, in many cases, the modeling of the investment in an asset as amount or share is irrelevant. However, there are cases where the use of amounts, instead of shares, is necessary.

3.2 The Structure of Transaction Costs

Let us indicate with $K(X_1, \ldots, X_n)$ the transaction cost function for a portfolio $X_1, \ldots, X_n$, that is the total transaction cost paid to buy an amount X_1 of asset 1, X_2 of asset 2, ... When a transaction cost is charged to an individual asset, we indicate with $K(X_j)$ the transaction cost to buy an amount X_j of asset j. In this book, we restrict the analysis to the most common case where transaction costs for the assets are independent from each other, that is to the case where the portfolio cost function is separable:

$$K(X_1, \ldots, X_n) = \sum_{j=1}^{n} K_j(X_j). \tag{3.1}$$

Fixed transaction costs A *fixed transaction cost* is a cost that is paid for handling an asset independently of the amount of money invested in it. The total transaction cost for asset j is expressed as:

$$K_j(X_j) = \begin{cases} f_j \text{ if } X_j > 0, \\ 0 \text{ otherwise,} \end{cases}$$

where f_j is the transaction cost paid if asset j is included in the portfolio.

We call the above cost structure imposing a fixed cost f_j for each selected asset j as *Pure Fixed Cost (PFC)*. This is one of the most common cost structures applied by financial institutions to investors. In the left-hand side of Fig. 3.1, we show a PFC function.

To include fixed costs in a portfolio optimization model we need binary variables z_j, one for each asset $j, j = 1, \ldots, n$. Variable z_j must be equal to 1 when asset j is selected in the portfolio, and to 0 otherwise. Constraints need to be added that

3.2 The Structure of Transcation Costs

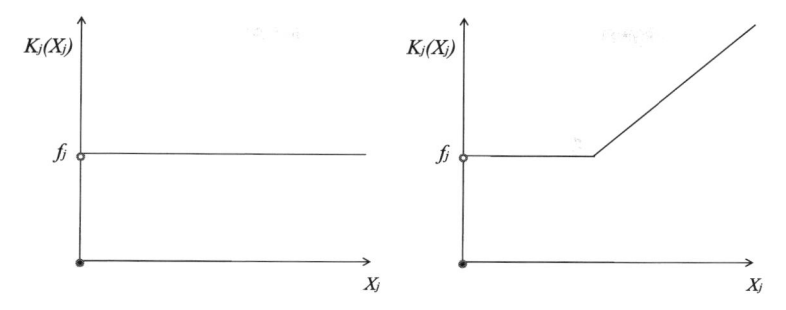

Fig. 3.1 Pure fixed cost structure (*left-hand side*), and proportional cost with minimum charge structure (*right-hand side*)

impose that z_j is equal to 1 if $X_j > 0$ and 0 otherwise. Thus, the following linear constraints must be added to any model:

$$L_j z_j \leq X_j \leq U_j z_j, \qquad j = 1, \ldots, n, \tag{3.2}$$

where L_j, U_j are positive lower and upper bounds on the amount invested in asset j, with U_j possibly equal to $\bar{C}$. The right-hand part of the constraint (3.2) imposes that if $X_j > 0$ then $z_j = 1$. The left-hand part forces $z_j = 0$ when $X_j = 0$. Note that, for a correct modeling, the lower bound L_j cannot be equal to 0. If no practical lower bound is defined, L_j may be chosen to be a very small real number.

With variable z_j defined as above, the transaction cost for asset j is expressed in linear form as

$$K_j(X_j) = f_j z_j.$$

We will see later different ways to account for the transaction costs in a portfolio optimization model.

Proportional transaction costs When the transaction cost depends on the amount invested we say that it is *variable*. Proportional transaction costs are the most frequently used in practice. A rate c_j is specified for each asset j, $j = 1, \ldots, n$, and the transaction cost is then expressed as a percentage of the invested amount in asset j, that is:

$$K_j(X_j) = c_j X_j. \tag{3.3}$$

We call this cost structure *Pure Proportional Cost (PPC)*.

Convex piecewise linear costs We consider here the case where each $K_j(X_j), j = 1, \ldots, n$, is a piecewise linear convex function. These functions are not frequently used in practice, but may be encountered to model some cost components as taxes.

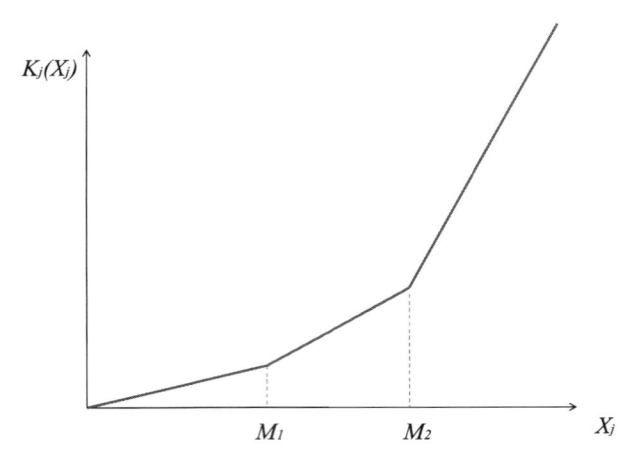

Fig. 3.2 A convex piecewise linear cost function

A different rate c_{ji} is applied to each, non-overlapping with others, interval $i \in I$ of capital invested in asset $j, j = 1, \ldots, n$, that is:

$$K_j(X_j) = \begin{cases} c_{j1} X_j & \text{if } 0 \leq X_j \leq M_1 \\ c_{j2}(X_j - M_1) + c_{j1} M_1 & \text{if } M_1 \leq X_j \leq M_2 \\ c_{j3}(X_j - M_2) + c_{j1} M_1 + c_{j2}(M_2 - M_1) & \text{if } M_2 \leq X_j \leq M_3 \\ \ldots \\ c_{j|I|}(X_j - M_{|I|-1}) + c_{j1} M_1 + c_{j2}(M_2 - M_1) \\ + \ldots + c_{j,|I|-1}(M_{|I|-1} - M_{|I|-2}) & \text{if } X_j \geq M_{|I|-1}, \end{cases} \qquad (3.4)$$

where $M_1, \ldots, M_{|I|-1}$ are the extremes defining the intervals into which the capital is divided. For the sake of simplicity, we assume that the extremes M_i do not depend on the asset. Note that, without loss of generality, we can always assume that also the last interval $(M_{|I|-1}, \infty)$ has a finite upper bound $M_{|I|}$ corresponding to the maximum amount that can be invested in an asset, i.e. the whole capital $\bar{C}$.

The rates c_{ji} are increasing, that is $c_{j1} < c_{j2} < c_{j3} < \ldots < c_{j|I|}$. In Fig. 3.2, we show an example of a convex piecewise linear function with three intervals.

Now, we show how to model these transaction costs in terms of variables and linear constraints. For each asset j we introduce one continuous variable X_{ji} for each interval $i \in I$, representing the capital invested in the interval. For example, consider asset j and the case of three intervals with $M_1 = 100$ and $M_2 = 500$. If the capital invested in asset j is 900, the X variables associated with asset j must take the values $X_{j1} = 100$, $X_{j2} = 400$ and $X_{j3} = 400$. Then, the investment X_j in asset j can be expressed as the sum of all auxiliary variables X_{ji} as:

$$X_j = \sum_{i \in I} X_{ji}. \qquad (3.5)$$

3.2 The Structure of Transaction Costs

We can express in linear form the relation between the X variables and the transaction cost as:

$$K_j(X_j) = \sum_{i \in I} c_{ji} X_{ji}. \tag{3.6}$$

In order to have a modeling with linear constraints we need to introduce for each asset j a binary variable z_{ji} for each interval i, $i = 2, \ldots, |I|$. The following constraints guarantee that $X_{j1} \leq M_1$, $X_{ji} \leq (M_i - M_{i-1})$ for $2 \leq i \leq |I| - 1$ and $X_{j,|I|} \leq (\bar{C} - M_{|I|-1})$. Moreover, they impose that if X_{ji} takes a positive value, all variables X_{jk}, $k = 1, \ldots, i-1$, are set to their upper bounds.

$$M_1 z_{j2} \leq X_{j1} \leq M_1 \tag{3.7a}$$

$$(M_2 - M_1) z_{j3} \leq X_{j2} \leq (M_2 - M_1) z_{j2} \tag{3.7b}$$

...

$$(M_{|I|-1} - M_{|I|-2}) z_{j,|I|} \leq X_{j,|I|-1} \leq (M_{|I|-1} - M_{|I|-2}) z_{j,|I|-1} \tag{3.7c}$$

$$0 \leq X_{j,|I|} \leq (\bar{C} - M_{|I|-1}) z_{j,|I|} \tag{3.7d}$$

It is worth noticing that, from a computational point of view, the use of binary variables could be avoided when the optimization model enforces minimal costs. In fact, due to the increasing cost rates, it would not be beneficial to activate variable X_{ji} unless all those with a smaller cost rate are saturated. Unfortunately, this is not always the case in portfolio optimization where a risk measure is minimized. Therefore, to represent this cost structure without relying on the optimization process, we should include constraints (3.7) imposing that, if variable X_{ji} takes a positive value, all variables X_{jk}, $k = 1, \ldots, i-1$, are forced to their upper bounds thanks to the introduction of binary variables.

Concave piecewise linear costs In this case each transaction cost function $K_j(X_j)$, $j = 1, \ldots, n$, is a piecewise linear concave function. The function is defined through (3.4). Here, however, the rates c_{ji} are decreasing, that is $c_{j1} > c_{j2} > c_{j3} > \ldots > c_{j|I|}$. This structure of transaction costs is commonly applied by financial institutions. In Fig. 3.3, we show an example of a concave piecewise linear function with 4 intervals, where $M_4 = \bar{C}$.

As for the convex piecewise linear cost functions, we introduce additional continuous variables X_{ji} and binary variables z_{ji}. Constraints (3.5) and (3.6), together with (3.7a)–(3.7d), complete the modeling of the transaction costs in linear form.

Another model can be built by using the list of all breaking points of the cost function. In addition to the r breaking points $(M_1, K_j(M_1)), \ldots, (M_r, K_j(M_r))$ we make use of the starting point $(M_0, K_j(M_0))$ corresponding to the origin $(0, 0)$. Again, we can assume that the last breaking point is $M_r = \bar{C}$. For example, in Fig. 3.3, we have four breaking points plus the origin.

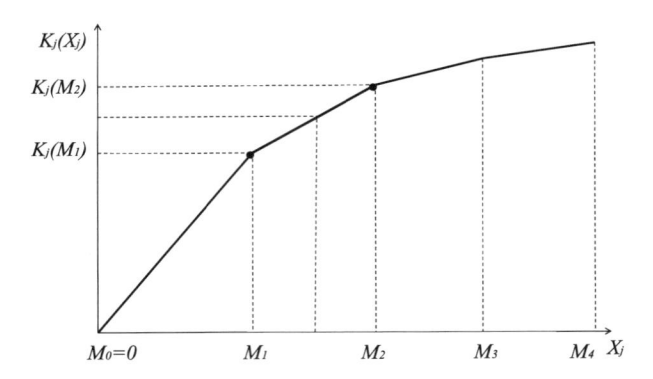

Fig. 3.3 A concave piecewise linear cost function

We know that each line segment can be expressed as the convex combination of its extremes. Therefore, if $M_i \leq X_j \leq M_{i+1}$, X_j can be expressed as $X_j = \lambda_{ji} M_i + \lambda_{j,i+1} M_{i+1}$ with $\lambda_{ji} + \lambda_{j,i+1} = 1$, $\lambda_{ji}, \lambda_{j,i+1} \geq 0$. Then, we obtain the transaction cost $K_j(X_j) = \lambda_{ji} K_j(M_i) + \lambda_{j,i+1} K_j(M_{i+1})$ (see, for instance, the line segment between $(M_1, K_j(M_1))$ and $(M_2, K_j(M_2))$ in Fig. 3.3).

To obtain constraints in linear form and a linear expression for the transaction costs, we need for each asset j a set of variables λ_{ji}, one for each breaking point $(M_i, K_j(M_i))$, $i = 0, \ldots, r$, and a set of binary variables z_{ji}, $i = 0, \ldots, r-1$, where $z_{ji} = 1$ if $M_i \leq X_j \leq M_{i+1}$, and $z_{ji} = 0$ otherwise. The linear constraints defining the meaning of the λ and z variables are as follows:

$$0 \leq \lambda_{j0} \leq z_{j0} \tag{3.8a}$$

$$0 \leq \lambda_{ji} \leq z_{j,i-1} + z_{ji} \qquad i = 1, \ldots, r-1 \tag{3.8b}$$

$$\lambda_{jr} \leq z_{j,r-1} \tag{3.8c}$$

$$\sum_{i=0}^{r-1} z_{ji} = 1. \tag{3.8d}$$

Thanks to these constraints, only one z_{ji} can take value 1. Moreover, for each j only pairs of adjacent coefficients λ_{ji} are allowed to be positive. Then, the variable X_j is defined as

$$X_j = \sum_{i=0}^{r} \lambda_{ji} M_i, \tag{3.9}$$

and the transaction cost takes the linear form

$$K_j(X_j) = \sum_{i=0}^{r} \lambda_{ji} K_j(M_i). \tag{3.10}$$

Linear costs with minimum charge A common piecewise linear cost function frequently used by financial institutions has the structure shown in Fig. 3.1 (right-hand side). This structure models a situation where a fixed cost f_j is charged for any invested amount lower than or equal to a given threshold M and then a proportional cost c_j is charged for an investment greater than M (we assume $M = f_j/c_j$):

$$K_j(X_j) = \begin{cases} 0 \text{ if } X_j = 0, \\ f_j \text{ if } 0 < X_j \leq M, \\ c_j X_j \text{ if } X_j > M. \end{cases}$$

We call this cost structure Proportional Cost with Minimum Charge (PCMC). This is a convex piecewise linear cost function with a discontinuity at $X_j = 0$.

To model this cost structure, we use the lower bound L_j on the amount invested in asset j. We introduce here two continuous variables X_{j1} and X_{j2} that have the same role of variables X_{ji} introduced earlier for the piecewise linear functions. Moreover, we make use of binary variables z_{j1} and z_{j2}. Then, we can model the transaction costs in linear form through the following constraints:

$$X_j = X_{j1} + X_{j2} \tag{3.11a}$$

$$L_j z_{j1} \leq X_{j1} \tag{3.11b}$$

$$M z_{j2} \leq X_{j1} \leq M z_{j1} \tag{3.11c}$$

$$0 \leq X_{j2} \leq (\bar{C} - M) z_{j2} \tag{3.11d}$$

with the cost formula:

$$K_j(X_j) = f_j z_{j1} + c_j X_{j2}. \tag{3.11e}$$

When $z_{j1} = z_{j2} = 0$ then $X_{j1} = X_{j2} = 0$, and $K_j(X_j) = 0$. When $z_{j1} = 1$ and $z_{j2} = 0$ then $L_j \leq X_{j1} \leq M$ and $X_{j2} = 0$, and $K_j(X_j) = f_j$. When $z_{j1} = z_{j2} = 1$ then $X_{j1} = M$ and $X_{j2} \leq \bar{C} - M$, and $K_j(X_j) = f_j + c_j(X_j - M) = f_j + c_j X_j - f_j = c_j X_j$. The case $z_{j1} = 0$ and $z_{j2} = 1$ is not feasible.

3.3 Accounting for Transaction Costs in Portfolio Optimization

We have seen how to express in linear form different structures of transaction costs. Now, the issue is how to include the transaction costs in a portfolio optimization model. Independently of the risk or safety function chosen as measure of the portfolio performance, a portfolio optimization model contains a constraint on the expected return of the portfolio

$$\sum_{j=1}^{n} \mu_j X_j \geq \mu_0 \bar{C},$$

and a constraint on the capital invested

$$\sum_{j=1}^{n} X_j = \bar{C}.$$

One may think about the transaction costs as reducing the capital available for the investment but also as diminishing the expected return. Moreover, the transaction costs may be accounted for in the objective function.

We will discuss the following different ways to account for transaction costs:

1. Costs treated separately: transaction costs may be considered as taken from a separate account. In this case an upper bound on the total cost is imposed in a separate constraint or, possibly, minimized in the objective function.
2. Costs deducted from the return: transaction costs are considered as a sum directly deducted from the expected portfolio return.
3. Costs deducted from the capital: transaction costs reduce the capital available. Thus, the invested capital is different from the initial capital being a variable that depends on the transaction costs paid.

Costs treated separately One may control the transaction costs by treating them separately, that is neither including them in the return nor in the capital constraints. This means adding to the model an additional constraint

$$\sum_{j=1}^{n} K_j(X_j) \leq K_{max},$$

where K_{max} is an upper bound on the total amount one is available to pay for transaction costs. The capital $\bar{C}$ is the invested capital and the return is $\sum_{j=1}^{n} \mu_j X_j$. A very large value of K_{max} corresponds to ignoring the transaction costs.

If this way of accounting for the transaction costs is chosen, one might explore the different portfolios generated by different values of K_{max}. Instead of bounding the costs one might decide to minimize them directly in the objective function.

Costs deducted from the return In this case, the constraint on the expected return of the portfolio is modified as follows:

$$\sum_{j=1}^{n} \mu_j X_j - \sum_{j=1}^{n} K_j(X_j) \geq \mu_0 \bar{C}, \qquad (3.12)$$

3.3 Accounting for Transaction Costs in Portfolio Optimization

that is the transaction costs diminish the average portfolio return. The capital available for the investment $\bar{C}$ is entirely used for the investment in the assets and is, thus, a constant. This models the situation where the transaction costs are charged at the end of the investment period.

A first crucial point here concerns the values of μ_j and μ_0. These values are assumed to be computed as return rates with the same time basis. For example, they may be all measured in €/month. This implies that the gross return $\sum_{j=1}^{n} \mu_j X_j$ is the amount of money earned after 1 month. Subtracting the transaction costs from a gross return calculated on a monthly basis corresponds to assuming that the investment horizon is 1 month. If instead the investment horizon is forecasted to be of 1 year, to account for the transaction costs correctly one must express the returns on a yearly basis, as €/year.

Given a fixed required return rate μ_0, including the transaction costs in this way implies that the gross return $\sum_{j=1}^{n} \mu_j X_j$ must be greater than it would be in the case without transaction costs. In other words, the higher the transaction costs the higher the gross return must be.

A second important point concerns the objective function and the way transaction costs deducted from the expected return may impact on portfolio performance. When we subtract the transaction costs from the return, we are focused on the net return $\bar{R}_X = R_X - K(X)$. Hence, in comparison to the original distribution of returns the net returns are represented with a distribution shifted by the transaction costs. Such a shift directly affects the expected value $\bar{\mu}(X) = \mathbb{E}\{\bar{R}_X\} = \mathbb{E}\{R_X\} - K(X) = \mu(X) - K(X)$. On the other hand, it does not affect the dispersion and $\bar{R}_X - \mathbb{E}\{\bar{R}_X\} = R_X - \mathbb{E}\{R_X\}$. Therefore, for dispersion type risk measures, like the variance, MAD, Semi-MAD, etc., we get for the net returns the same value of risk measure as for the original returns. This means that such a risk measure minimization is neutral with respect to the transaction costs. Therefore, when using dispersion type risk measure minimization an investor may face difficulties related to discontinuous efficient frontier and non-unique, possibly inefficient, optimal solution. Essentially, the problem of non-unique minimum risk portfolios with bounded required return may arise also when transaction costs are ignored. As a very naive example one may consider a case where two risk-free assets are available with different returns but both meeting the required expected return. In the following, we show a non-trivial example of non-unique solution for the case of PCMC transaction costs.

Example 3.1 Let us consider the case of proportional transaction costs with minimum charge (see Fig. 3.1, right-hand side). Let us assume that the minimum charge is $f_j = 50$ €, whereas the proportional cost is on the level of 1 %, i.e. $c_j = 0.01$, and

$$K_j(X_j) = \begin{cases} 0 & \text{if } X_j = 0 \\ \max\{50, 0.01 X_j\} & \text{if } X_j > 0. \end{cases}$$

Consider 3 assets with rates of return under 3 equally probable scenarios $t = 1, 2, 3$, as presented in Table 3.1. The capital to be invested is $\bar{C} = 10{,}000$ €.

Table 3.1 Rate of returns for three assets under three scenarios

Asset	Scenario 1 (%)	Scenario 2 (%)	Scenario 3 (%)	Mean return rate (%)
1	$r_{11} = 14.67$	$r_{12} = 14.67$	$r_{13} = 17.67$	$\mu_1 = 15.67$
2	$r_{21} = 15.07$	$r_{22} = 16.07$	$r_{23} = 14.07$	$\mu_2 = 15.07$
3	$r_{31} = 15.92$	$r_{32} = 14.92$	$r_{33} = 13.92$	$\mu_3 = 14.92$

Consider a portfolio (X_1, X_2, X_3) in terms of amounts defining a portfolio, $X_1 + X_2 + X_3 = \bar{C}$. The expected net return of the portfolio taking into account the transaction costs is given by the following formula

$$\bar{\mu}(\mathbf{X}) = (0.1567X_1 + 0.1507X_2 + 0.1492X_3) - K_1(X_1) - K_2(X_2) - K_3(X_3).$$

Assume we are looking for minimum risk portfolio meeting the specified lower bound on the expected net return, say 14 %, i.e. $\bar{\mu}(\mathbf{X}) \geq 0.14\bar{C}$.

We focus the analysis on the risk measured by the Semi-MAD

$$\bar{\delta}(X) = \sum_{t=1}^{T} p_t \max\{\mu(X) - \sum_{j=1}^{n} r_{jt}X_j, 0\}.$$

Note that the absolute risk measure is defined on distributions of returns not reduced by the transaction costs. However, the transaction costs are deterministic, which causes that the same risk measure remains valid for the net returns.

Finally, the problem of minimizing risk under the required return takes the following form:

$$\min\{\bar{\delta}(X) : X_1 + X_2 + X_3 = \bar{C}, X_j \geq 0 \ \forall j, \bar{\mu}(\mathbf{X}) \geq 0.14\bar{C}\} \tag{3.13}$$

One can find out that the required return bound may be satisfied by: two single asset portfolios $(\bar{C}, 0, 0)$ and $(0, \bar{C}, 0)$ as well as by some two-asset portfolios with large enough share of the first asset, i.e., portfolios $(X_1, X_2, 0)$ with $X_1 \geq 43/160\bar{C}$, and portfolios $(X_1, 0, X_3)$ with $X_1 \geq 58/175\bar{C}$. On the other hand, no two-asset portfolio built using assets 2 and 3, and no three-asset portfolio fulfills the required return bound.

While minimizing the Semi-MAD measure one can see that $\bar{\delta}(\bar{C}, 0, 0) = 0.04\bar{C}/3$ and $\bar{\delta}(0, \bar{C}, 0) = 0.02\bar{C}/3$. Among portfolios $(X_1, X_2, 0)$ the minimum Semi-MAD measure is achieved for $\bar{\delta}(\bar{C}/3, 2\bar{C}/3, 0) = 0.02\bar{C}/9$. Similarly, among portfolios $(X_1, 0, X_3)$ the minimum Semi-MAD measure is achieved for $\bar{\delta}(\bar{C}/3, 0, 2\bar{C}/3) = 0.02\bar{C}/9$. Thus, finally, while minimizing the Semi-MAD measure one has two alternative optimal portfolios $(\bar{C}/3, 2\bar{C}/3, 0)$ and $(\bar{C}/3, 0, 2\bar{C}/3)$. Both portfolios have the same Semi-MAD value but they are quite different with respect to the expected return. Expected return of portfolio $(\bar{C}/3, 0, 2\bar{C}/3)$ is smaller

3.3 Accounting for Transaction Costs in Portfolio Optimization

than that of portfolio $(\bar{C}/3, 2\bar{C}/3, 0)$. Therefore, portfolio $(\bar{C}/3, 0, 2\bar{C}/3)$ is clearly dominated (inefficient) in terms of mean-risk analysis. While solving problem (3.13) one may receive any of two minimum risk portfolios. It may be just the dominated one. Therefore, the corresponding safety measure maximization must be used to guarantee selection of an efficient portfolio

$$\max\{\bar{\mu}(X) - \bar{\delta}(X) : X_1 + X_2 + X_3 = \bar{C}, X_j \geq 0 \ \forall j, \bar{\mu}(X) \geq 0.14\bar{C}\}. \quad (3.14)$$

Costs deducted from the capital To account for the transaction costs in the capital constraint, the capital invested in the assets becomes

$$C = \bar{C} - \sum_{j=1}^{n} K_j(X_j),$$

with

$$\sum_{j=1}^{n} X_j = C,$$

which implies that the invested capital C is a variable, in general different from the initial available capital $\bar{C}$, depending on the amount of transaction costs paid.

Reducing the capital available corresponds to assuming that the transaction costs are charged at the time the investment in the assets is made. Actually, decreasing the invested capital, the transaction costs decrease also indirectly the (net) return. The return is a difference between the current wealth (portfolio value) and the initial capital $\bar{C}$. Hence, the expected return can be expressed as

$$\bar{\mu}(X) = C + \sum_{j=1}^{n} \mu_j X_j - \bar{C} = \sum_{j=1}^{n} \mu_j X_j - \sum_{j=1}^{n} K_j(X_j)$$

and the constraint on the expected return of the portfolio takes the same form

$$\sum_{j=1}^{n} \mu_j X_j - \sum_{j=1}^{n} K_j(X_j) \geq \mu_0 \bar{C}, \quad (3.15)$$

as for transaction costs deducted from return. In (3.15) the required expected return is computed with respect to the available capital $\bar{C}$. To impose a required expected return with respect to the invested capital, $\bar{C}$ should be replaced by C. Clearly, in the latter case, the right-hand side of (3.15) is smaller than in the former case and has a different interpretation.

Table 3.2 Transaction costs in the capital constraint: an example

Asset	Scenario 1 (%)	Scenario 2 (%)	Mean return rate (%)	Fixed transaction cost (€)
1	$r_{11} = 2$	$r_{12} = 4$	$\mu_1 = 3$	4
2	$r_{21} = 2$	$r_{22} = 4$	$\mu_2 = 3$	4

The impact of this way of accounting for the transaction costs in a portfolio optimization model is that the capital available for the investment in the assets is reduced by the transaction costs. Being an amount, the higher the transaction costs the lower the invested capital. This implies a somehow surprising and unpleasant consequence. When a convex risk function is chosen, a minimization of the risk function pushes the optimization toward an increase of the transaction costs because this reduces the invested capital and, thus, the risk. In other words, to reduce the risk we aim at reducing the money invested and thus, as the capital available is given, at increasing the transaction costs. This kind of portfolio would not make any investor happy. The problem does not take place when a safety measure is used instead.

Example 3.2 In Table 3.2 we show an example of $n = 2$ assets and $T = 2$ scenarios with each scenario having probability 50 %. The table shows the rates of return of the assets in the different scenarios, the average return rate and the fixed transaction cost paid for the inclusion of the asset in the portfolio. Suppose that a capital of 1,000 € is available and the required return is equal to $\mu_0 = 2\%$.

As the assets are identical any investor would invest all the capital in one only of the two assets, say asset 1, paying 4 € of transaction cost and investing 996 € in the asset. In this case, the average return of the investment is $\mu(X_1) = \sum_{s=1}^{2} p_s r_{1s} X_1 = 0.5 * 0.02 * 996 + 0.5 * 0.04 * 996 = 29.88$ and the net return $\bar{\mu}(X_1) = 29.88 - 4 = 25.88$. If the Semi-MAD is used as risk function, then the risk corresponding to the investment of 996 € in asset 1 is $\sum_{t=1}^{2} p_t \max\{29.88 - r_{1t}X_1, 0\} = 0.5(29.88 - 0.02 * 996) + 0.5 * 0 = 4.98$. Note that the net return exceeds the required return by $25.88 - 20.0 = 5.88$ €.

The Semi-MAD optimal portfolio will instead select both assets in the portfolio. Paying 8 € transaction costs, the capital invested becomes 992 €. Then 991 € are invested in asset 1 and 1 € in asset 2, the mean return of the portfolio is $\sum_{s=1}^{2} p_s \sum_{j=1}^{2} r_{js} X_j = 0.5(0.02*991+0.02*1) + 0.5(0.04*991+0.04*1) = 29.76$ and the net return $\bar{\mu}(X) = 29.76 - 8 = 21.76$ while the risk is $\sum_{t=1}^{2} p_t \max\{29.76 - \sum_{j=1}^{2} r_{jt}X_j\} = 0.5(29.76 - 0.02 * 991 - 0.02 * 1) + 0.5 * 0 = 4.96$. Note that the net return exceeds the required return by $21.76 - 20.0 = 1.76$ €.

Thus, investing in two identical assets, and paying a fixed transaction cost for each, reduces the risk with respect to investing in one asset only and paying the transaction cost for that asset only. Note that while focusing on the corresponding Semi-MAD safety measure one gets a reasonable result of preferred single asset portfolio with measure $\bar{\mu}(X) - \bar{\delta}(X) = 25.88 - 4.98 = 20.9 > 16.8 = 21.76 - 4.96$.

3.4 Optimization with Transaction Costs

In Sect. 3.2, we have seen different structures of transaction costs and how in each of these cases the transaction costs can be expressed in linear form. In this section, we see that it is not always necessary to include in a portfolio optimization model all the constraints presented to express the transaction costs. In fact, there are cases where the optimization model *supports cost minimization*. In such cases, it is sufficient to have constraints imposing that the transaction costs must be greater or equal than the exact value. Then, the objective function will push the transaction costs to the minimum value allowed by the constraints.

Actually, this is the case when safety forms of risk measures are used. Optimization based on such measures supports cost minimization and therefore allows us for simpler models while avoiding inconsistencies discussed in the previous section.

Convex piecewise linear costs This case is easy to model in terms of linear constraints and continuous variables when the costs are minimized. Namely, introducing one continuous variable X_{ji} for each interval $i \in I$, the investment X_j in asset j can be expressed as the sum of all auxiliary variables X_{ji} as:

$$X_j = \sum_{i \in I} X_{ji} \tag{3.16a}$$

and the transaction cost function as:

$$K_j(X_j) = \sum_{i \in I} c_{ji} X_{ji} \tag{3.16b}$$

with the additional constraints:

$$0 \leq X_{j1} \leq M_1 \tag{3.16c}$$

$$0 \leq X_{j2} \leq (M_2 - M_1) \tag{3.16d}$$

$$\ldots$$

$$0 \leq X_{j,|I|-1} \leq (M_{|I|-1} - M_{|I|-2}) \tag{3.16e}$$

$$0 \leq X_{j,|I|} \leq \bar{C}. \tag{3.16f}$$

These constraints guarantee only that the transaction costs will not be lower than $K_j(X_j)$. However, they are sufficient when relying on optimization models to enforce minimal costs. In fact, due to the increasing cost rates, an optimization model that supports cost minimization will not find beneficial to activate variable X_{ji} unless all those with a smaller cost rate are saturated.

Concave piecewise linear costs Contrary to the case of convex functions, the variable X_{ji} has a smaller rate and, thus, is more convenient than all variables X_{jk}, $k = 1, \ldots, i-1$ associated with the preceding intervals. Therefore, even in optimization models that support cost minimization we need the full set of constraints defining the transaction costs exactly.

Linear costs with minimum charge In optimization models that support cost minimization the condition that transaction costs must be not lower than the real value can be simply expressed as

$$k_j \geq f_j z_j, \quad k_j \geq c_j X_j, \quad 0 \leq X_j \leq \bar{C} z_j, \tag{3.17}$$

where z_j are binary variables and $k_j = K_j(X_j)$.

3.5 A Complete Model with Transaction Costs

In this section, we provide a complete formulation of a portfolio optimization problem with transaction costs. The objective function is the CVaR as defined by (2.20) in Chap. 2 which has become a very popular safety measure. We consider the given tolerance level $\beta > 0$. As cost function $K_j(X_j)$ for each asset j, $j = 1, \ldots, n$, we introduce both a proportional and a fixed cost. The transaction costs are deducted from the capital, and therefore from the mean return too.

$$\max \eta - \frac{1}{\beta} \sum_{t=1}^{T} p_t d_t \tag{3.18a}$$

$$\eta - \sum_{j=1}^{n}(r_{jt} - c_j)X_j + \sum_{j=1}^{n} f_j z_j \leq d_t \qquad t = 1, \ldots, T \tag{3.18b}$$

$$\sum_{j=1}^{n}(\mu_j - c_j)X_j - \sum_{j=1}^{n} f_j z_j \geq \mu_0 \bar{C} \tag{3.18c}$$

$$L_j z_j \leq X_j \leq U_j z_j \qquad j = 1, \ldots, n \tag{3.18d}$$

$$\sum_{j=1}^{n} X_j + \sum_{j=1}^{n} c_j X_j + \sum_{j=1}^{n} f_j z_j = \bar{C} \tag{3.18e}$$

$$d_t \geq 0 \qquad t = 1, \ldots, T \tag{3.18f}$$

$$X_j \geq 0 \qquad j = 1, \ldots, n \tag{3.18g}$$

$$z_j \in \{0, 1\} \qquad j = 1, \ldots, n. \tag{3.18h}$$

Variable η is the independent free variable which at optimality represents the value of the β-quantile. Binary variable z_j, $j = 1,\ldots,n$, is set equal to 1 when the corresponding asset is selected in the portfolio and 0 otherwise.

Constraints (3.18b) along with constraints (3.18f) define each variable d_t as the $\max\{0, \eta - \eta_t\}$, where $\eta_t = \sum_{j=1}^{n}(r_{jt} - c_j)q_j X_j - \sum_{j=1}^{n} f_j z_j$ is the portfolio realization under scenario t when fixed and proportional costs are taken into account (net portfolio realization). The transaction cost function for each asset j is thus $K_j(X_j) = c_j X_j + f_j z_j$. Constraint (3.18c) imposes that the net portfolio mean return has to be greater than or equal to the required return μ_0 applied to the capital amount $\bar{C}$. Constraints (3.18d) define the lower and upper bounds on the investment for asset j equal to L_j and U_j, respectively. Constraint (3.18e) defines balance of the invested capital and transaction costs with the available capital. Finally, constraints (3.18f)–(3.18h) are non-negative and binary conditions.

We recall that the use of a safety measure is essential to avoid the previously mentioned unpleasant consequences of the introduction of transaction costs.

3.6 Notes and References

In the literature the number of contributions that explicitly account for transaction costs in portfolio optimization is quite limited. Such a number further reduces if only linear and mixed integer linear programming problems are taken into account. A general survey on transaction costs along with other real features in mixed integer linear programming problems is provided in Mansini et al. (2014), whereas a specific focus on transaction costs can be found in Mansini et al. (2015). For a survey on transaction costs in a mean-variance framework we refer to the recent work by Chen et al. (2010).

Speranza (1996) is one of the first papers directly accounting for fixed transaction costs. Also Young (1998) extended his linear Minimax model to include for each asset both a variable cost for unit purchased and a fixed one. He also underlined the importance of the investment horizon when computing the net expected return specifying the number of periods in which the portfolio would be held. Different mixed integer linear programming models using the Semi-MAD as risk measure and dealing with different structures for fixed costs can be found in Kellerer et al. (2000). This paper provides the proof that finding a feasible solution to the portfolio optimization problem with fixed costs is NP-complete. Other mixed integer linear programming models dealing with fixed transaction costs can be found in Angelelli et al. (2008, 2012), Baumann and Trautmann (2013), Chiodi et al. (2003), Guastaroba et al. (2009a), and Mansini and Speranza (2005).

Variable costs are more commonly treated in the literature. We refer to Chiodi et al. (2003) for an example of a portfolio problem with mutual funds including concave piecewise linear function to model entering commissions. A structure with stepwise increasing transaction costs can be found in Le Thi et al. (2009), whereas Konno and Yamamoto (2005) considered a portfolio optimization problem where transaction cost functions are piecewise linear concave and piecewise constant. Konno and Wijayanayake (2001) analyzed a portfolio construction/rebalancing problem under concave transaction costs.

For an analysis of the different modeling alternatives to include the transaction costs in a portfolio optimization model we refer to Angelelli et al. (2008), Angelelli et al. (2012), Chiodi et al. (2003), Kellerer et al. (2000), Konno et al. (2005), Konno and Yamamoto (2005), Krejić et al. (2011), Lobo et al. (2007), and Mansini and Speranza (2005) to find examples of portfolio optimization models where transaction costs (fixed or variable) were deducted from the return. Examples of models where transactions costs reduced the capital available for the investment and/or the costs were upper bounded inside a specific constraint were introduced in Beasley et al. (2003), Lobo et al. (2007), Woodside-Oriakhi et al. (2013), and Young (1998).

Chapter 4
Portfolio Optimization with Other Real Features

4.1 Introduction

Transaction costs represent the most important feature to account for when selecting a real portfolio. However, it is not the only one. In real applications, an investor may be obliged to limit the amount invested in an asset or may wish to control the number of assets selected. More generally, *real features* are all the additional characteristics an investor is interested to consider, because they reflect preferences or information not captured by the model otherwise, or he/she is obliged to include as restrictions imposed by market conditions. Besides transaction costs, real features include, for example, transaction lots, thresholds on investment, limitation on the number of assets (cardinality constraint), and decision dependency constraints among assets or classes of assets.

In this chapter, we focus on real features different from transaction costs. We analyze their practical relevance and show how to model them in a portfolio optimization problem. The modeling of some real features is possible by using as decision variables the asset shares (variables x_j, $j = 1, \ldots, n$). In several cases, however, the introduction of real features implies the need of variables that represent the amount of capital invested in each asset (variables X_j, $j = 1, \ldots, n$). When discussing each real feature, and the way to treat it in a portfolio optimization model, we will specify if the use of amounts instead of shares is required.

As in Chap. 3, we refer to the available capital as $\bar{C}$, and indicate the invested capital, when it does not coincide with the available capital, as variable C.

4.2 Transaction Lots

A *transaction lot*, also simply called *lot* or *round*, represents a standardized quantity associated with a specific asset, set by a regulatory body. A lot represents the basis for a transaction on the asset. For exchange-traded assets, a lot represents the minimum quantity of an asset that can be traded. Stocks are usually traded in units and thus, in terms of stocks, the lot is the minimum number of units of a stock that can be purchased in one transaction. The concept of lot allows the financial markets to standardize quotations. For example, options are priced in such a way that each contract (or lot) represents exercise rights for 100 underlying units of a stock. With such standardization, investors always know exactly how many units they are buying with each contract and can easily assess what price per unit they are paying. Without such standardization, valuing and trading options would be needlessly cumbersome and time consuming. In many stock exchanges, the transaction lot for a stock corresponds to a single stock unit. An investor may thus decide to buy 50 units of a stock, investing an amount of capital equal to the stock quotation (the monetary value of one unit of the stock) multiplied by 50.

For the sake of a simple presentation, we will stretch the concept of lot in this chapter. We will say that we buy a certain number of lots of an asset and we will call *lot value* the minimum amount of capital that can be invested in that asset, that is the monetary value of the standardized quantity for that asset. Besides, the capital invested in the asset has to be expressed as a multiple of such a transaction lot value. For instance, if an asset has a standardized lot of 15 units and the current quotation of a single unit is equal to 1.5 €, then the monetary counter value of a lot for that asset is equal to 22.5 €. In this case, we will say that the lot value is 22.5, and if we buy 7 lots of that asset, we will invest a capital equal to $7 * 22.5 = 157.5$ €.

When considering transaction lots, the capital made available for the investment becomes a critical decision. Let us consider the following example:

Example 4.1 An investor has to decide how much to invest in each of two assets. Transaction lot values are equal to 1.73 and 9.50 € for the first and the second asset, respectively. The investor wishes to invest 100 €. It is easy to see that no feasible portfolio can be found corresponding to exactly such an amount. If, however, we allow the capital to range between 100 and 101 €, then a feasible portfolio can be found: we buy 3 lots of the first asset and 10 lots of the second one. This implies an amount equal to 5.19 € invested in the first asset and 95 € in the second one, for a total capital invested equal to 100.19 €.

Thus, in order to guarantee that the portfolio optimization can find a feasible solution the capital invested cannot be restricted to be a fixed amount $\bar{C}$ but must be a variable C ranging between a minimum and a maximum value as follows:

$$\bar{C}_L \leq C \leq \bar{C}_U,$$
C continuous,

4.2 Transaction Lots

where $\bar{C}_L$ and $\bar{C}_U$ represent the minimum and maximum amount of money the investor is available to invest. Whereas the reason to have the upper bound $\bar{C}_U$ is obvious, one may think that the lower bound $\bar{C}_L$ is not necessary and that without it the model will select the optimum amount of capital to invest. In fact, without the lower bound the optimization of a risk function tends to invest the minimum amount of capital that guarantees the required expected return of the investment. With investments expressed as amounts and risk measures, the smaller the capital invested the smaller the risk. With safety measures, the lower bound is not necessary.

A second important aspect to consider when introducing transaction lots in an optimization model is the type of variables used to represent investments.

Let V_j represent the monetary value of the transaction lot for asset j and let χ_j be the nonnegative integer variable representing the number of lots of asset j selected in the portfolio. We now consider separately how to model transaction lots using amounts invested or, as an alternative, shares.

1. **The case of amounts.** In optimization models considering variables $X_j, j = 1, \ldots, n$, as amounts invested, the transaction lot for asset j can be modeled as:

$$X_j = V_j \chi_j \quad (4.1)$$

$$\chi_j \geq 0 \quad \text{integer} \quad (4.2)$$

$$X_j \geq 0 \quad (4.3)$$

with $\sum_{j=1}^{n} X_j = C$.

2. **The case of shares.** In optimization models using shares $x_j, j = 1, \ldots, n$, the formulation for transaction lots on asset j becomes:

$$C x_j = V_j \chi_j \quad (4.4)$$

$$\chi_j \geq 0 \quad \text{integer} \quad (4.5)$$

$$0 \leq x_j \leq 1, \quad (4.6)$$

with $\sum_{j=1}^{n} x_j = 1$.

Note that constraint (4.4) introduces a nonlinear relation. There are two ways to avoid the quadratic expression $C x_j$. The simplest approach is to use variables that represent amounts invested as in case 1 analyzed above. As an alternative, one may consider amounts invested relatively to the capital available $\bar{C}_U$ rather than to the capital invested C, leading to the following formulation for asset j:

$$\bar{C}_U x_j = V_j \chi_j \quad (4.7)$$

$$\chi_j \geq 0 \quad \text{integer} \quad (4.8)$$

$$0 \leq x_j \leq 1, \quad (4.9)$$

Table 4.1 The impact of transaction lots: an example

Asset	Scenario 1 (%)	Scenario 2 (%)	Mean return rate (%)	Transaction lot value (€)
1	$r_{11} = 3$	$r_{12} = 5$	$\mu_1 = 4$	12
2	$r_{21} = 0$	$r_{22} = 10$	$\mu_2 = 5$	10
3	$r_{31} = 3$	$r_{32} = 3$	$\mu_3 = 3$	40

with

$$\sum_{j=1}^{n} x_j \leq 1 \tag{4.10}$$

$$C = \bar{C}_U \sum_{j=1}^{n} x_j. \tag{4.11}$$

The invested amounts corresponding to the shares $x_j, j = 1, \ldots, n$, are now available as quantities $\bar{C}_U x_j$ and the model has linear constraints. In this case, the quantity $1 - \sum_{j=1}^{n} x_j$ represents the percentage of non-invested capital calculated with respect to $\bar{C}_U$.

In the literature and in practice, the transaction lots are frequently ignored. However, in this case, solutions may turn out to be not implementable, as shown in the following example.

Example 4.2 We have two scenarios each having probability 50 %, and three assets with transaction lot values equal to 12, 10 and 40 €, respectively. Table 4.1 shows the rates of return (in percentage) of the assets in the different scenarios, the mean return rate and the transaction lot value of each asset. The capital available for the investment can range between $\bar{C}_L = 100\,€$ and $\bar{C}_U = 102\,€$, whereas the required rate of return is $\mu_0 = 3\,\%$. The portfolio performance is measured by the Semi-MAD. Ignoring transaction lots would lead to an optimal solution where only asset 3 (the risk free asset) is selected with the minimum possible amount of capital, 100 €, invested. If, on the contrary, we consider transaction lots, then the optimal solution would be to invest 102 € by selecting 2 lots of asset 3 and 1 lot of assets 1 and 2. In fact, with 2 lots of asset 3 we get an investment of 80 € and we do not reach the lower bound $\bar{C}_L = 100$, but with 3 lots we exceed the maximum investment allowed. However, adding 1 lot of asset 1 to the portfolio brings the investment to 92 €, that does not reach the minimum investment, whereas 2 lots of asset 1 exceeds the maximum investment.

Looking at the previous two examples we can make two remarks. The first one is that, in general, if transaction lots are ignored the optimization does not identify the set of assets contained in the optimal portfolio when lots are considered. The second aspect concerns the total amount invested. As for all risk measures, the Semi-MAD pushes the capital invested to the minimum value that satisfies the constraints and in

4.3 Thresholds on Investment

Table 4.2 Risk minimization and capital invested: an example

Asset	Scenario 1 (%)	Scenario 1 (%)	Mean return rate (%)	Transaction lot value (€)
1	$r_{11} = 2$	$r_{12} = 4$	$\mu_1 = 3$	3
2	$r_{21} = 4$	$r_{22} = 6$	$\mu_2 = 5$	6

particular the required expected return. In Example 4.1 this behaviour is illustrated, whereas in Example 4.2 the unique feasible solution requires to invest a capital equal to the upper bound.

It may also happen that, to minimize the risk, the model will choose to invest less capital possibly missing an additional return, as shown by the following example.

Example 4.3 Let us consider the two assets in Table 4.2, with $T = 2$ scenarios, each having probability 50 %. The range of the capital $[\bar{C}_L, \bar{C}_U]$ is set to [99, 102], whereas the required expected return is equal to $\mu_0 = 2.5\%$ and transaction lot values are equal to 3 and 6 €, respectively. Risk is measured by the Semi-MAD. Note that the first asset carries the same individual risk of the second asset but brings lower return, whereas both assets have mean rates of return higher than the required one. In order to minimize the risk, the model will try to reduce as much as possible the capital invested and, thus, will select the first asset that, due to the lower value of the transaction lot, allows an investment equal to $\bar{C}_L$. The optimal solution is then $\chi_1 = 33$ and $\chi_2 = 0$ for an investment equal to 99 €. The portfolio rate of return and risk are equal to 2.97 % and 0.495, respectively. Note that the solution selecting asset 2 only ($\chi_1 = 0$, $\chi_2 = 17$) is also feasible, but would require an investment equal to 102 € with portfolio rate of return equal to 5.1 % and risk equal to 0.51.

To avoid the problem observed in this example one should consider a safety measure, rather than a risk measure.

4.3 Thresholds on Investment

Another type of practical limitation consists in imposing lower and upper limits on the share or on the amount of an asset held in the portfolio. These constraints are commonly identified as *threshold constraints*. These requirements come from real-world practice where a large number of assets with very small holdings is not desirable.

Thus, on asset j, the constraint that imposes lower and upper bounds is formulated as:

$$l_j \leq x_j \leq u_j \quad (4.12)$$

in models with shares, and as

$$L_j \leq X_j \leq U_j \tag{4.13}$$

in models with amounts, where l_j (u_j) and L_j (U_j) are the lower (upper) bounds on the investment in asset j, the former expressed in percentage, the latter in amount of capital. Notice that if $l_j > 0$ ($L_j > 0$), then the selection of asset j is forced in the portfolio.

If the lower and upper bounds are intended to hold only if the asset is selected in the portfolio, the above constraints must be replaced by constraints that make use of the binary variables $z_j, j = 1, \ldots, n$, where z_j is equal to 1 if asset j is selected and 0 otherwise:

$$l_j z_j \leq x_j \leq u_j z_j \tag{4.14}$$

if the model uses shares, and

$$L_j z_j \leq X_j \leq U_j z_j \tag{4.15}$$

if the model considers amounts. Possible trivial values for the upper bounds are $u_j = 1$ and $U_j = \bar{C}$, respectively.

In a number of applications it is also relevant to have the possibility to set lower and upper bounds on the investment in groups of assets (the so called *class constraints*). An asset class is a group of assets sharing similar characteristics, such as riskiness and return. Different asset classes may offer returns that are loosely correlated, whereas correlation within the same class is usually high. Hence, diversification over the classes reduces the overall risk of the portfolio.

Basic asset classes correspond to stocks, bonds, and cash. However, inside each of these classes, one may consider groups of assets. For instance, assets may be divided according to the sector/industry they belong to. Class constraints may aim to limit the investment in a specific sector of assets considered extremely risky, through an upper bound, or to guarantee a minimum investment in a promising area of investment, through a lower bound. Sectors include commodities (metals, agriculture, energy), banking and insurance, real estate, government, high tech, manufacturing, just to name a few.

Let G_s be a set of assets of the same sector s. A class constraint is formulated, in models with shares, as follows:

$$l_s \leq \sum_{j \in G_s} x_j \leq u_s, \tag{4.16}$$

where l_s and u_s are lower and upper bounds expressed as percentages of the total unitary investment. Similarly, in the case of models with amounts invested, with the X's instead of the x's and the constants that represent amounts.

4.3 Thresholds on Investment

Finally, there is another important issue that concerns the selection of quite undiversified portfolios when using safety instead of risk measures. Safety measures tend to select undiversified portfolios. Although this result is theoretically justified, an investor may be concerned about the quality of the data and wish to impose additional constraints on the model possibly forcing a minimum portfolio diversification.

The simplest way to obtain such a result is to limit the maximum share (maximum amount) as in (4.12) and (4.13). This, however, may result in a portfolio with a few equal shares depending on the value set on the maximum share. A better modeling alternative would be to allow for a relatively large maximum share provided that the other shares are smaller. Such a rich diversification scheme may be introduced applying the CVaR concept to the right tail of the distribution of portfolio shares.

More precisely, a natural generalization of the maximum share is the (right-tail) conditional mean defined as the mean within the specified tolerance level (amount) of the worst shares. One may simply define the conditional mean as the mean of the k largest shares through the solution of the linear problem:

$$\min\{ks_k + \sum_{j=1}^{n} d_{kj}^s : d_{kj}^s \geq x_j - s_k, d_{kj}^s \geq 0 \quad j = 1, \ldots, n\},$$

where s_k is an unbounded variable (representing the k-th largest share at the optimum) and d_{kj}^s are additional nonnegative (deviational) variables.

Hence, any model can be extended with direct *diversification enforcement constraints* upper bounding the value of the k largest shares by a percentage γ_k, implemented by the linear inequalities:

$$ks_k + \sum_{j=1}^{n} d_{kj}^s \leq \gamma_k \text{ and } d_{kj}^s \geq x_j - s_k, d_{kj}^s \geq 0 \quad j = 1, \ldots, n. \qquad (4.17)$$

Example 4.4 Let us consider the following diversification conditions. We want to impose that in the selected portfolio any asset share cannot exceed 0.25, while any three shares cannot exceed 0.60 in total and any six shares cannot globally exceed 0.80 of the portfolio investment. This requires the introduction of the following constraints:

$$x_j \leq 0.25 \qquad\qquad\qquad\qquad j = 1, \ldots, n \qquad (4.18)$$

$$3s_3 + \sum_{j=1}^{n} d_{3j}^s \leq 0.60, \, d_{3j}^s \geq x_j - s_3, \, d_{3j}^s \geq 0 \qquad j = 1, \ldots, n \qquad (4.19)$$

$$6s_6 + \sum_{j=1}^{n} d_{6j}^s \leq 0.80, \, d_{6j}^s \geq x_j - s_6, \, d_{6j}^s \geq 0 \qquad j = 1, \ldots, n. \qquad (4.20)$$

4.4 Cardinality Constraints

One of the basic implications of modern portfolio theory is that investors prefer to hold well diversified portfolios to minimize risk. However, in practice, investors typically prefer portfolios including only a limited number of assets. We have already shown how market imperfections, as fixed transaction costs, are related to the number of assets in a portfolio. Frequently, the need to avoid such costs as well as the cost of monitoring and/or re-balancing portfolios lead investors to select scarcely diversified portfolio. A restriction on the number of assets that can be selected in a portfolio is called *cardinality constraint*.

To incorporate the cardinality constraint in a portfolio optimization model we need to use binary variables z_j, $j = 1, \ldots, n$, where z_j is equal to 1 if asset j is selected in the portfolio, and to 0 otherwise. Then, the cardinality constraint can be expressed imposing that the number of selected assets cannot be greater than a predefined number K_{sup} and not lower than a minimum number K_{inf}:

$$K_{inf} \leq \sum_{j=1}^{n} z_j \leq K_{sup}. \tag{4.21}$$

To correctly enforce the value of binary variables, this constraint is usually associated with the threshold constraints (4.14) if model uses shares or constraints (4.15) in case of amounts.

In portfolio optimization the cardinality constraint (4.21) is often imposed as a strict equality, with $K_{sup} = K_{inf} = K$. Inequalities are frequently used in index tracking problems, where the target of the investor is to replicate a benchmark index without investing in all the assets that compose it.

It is interesting to note that, if we know the minimum positive share size for each asset ($\varepsilon > 0$), then the lower limit on portfolio cardinality can be easily implemented using the diversification enforcement constraints as follows.

The sum of k largest shares in a portfolio $\mathbf{x}$ may be expressed by the linear program:

$$\min \{k\eta + \sum_{j=1}^{n} s_j \ : \ s_j \geq x_j - \eta, \ s_j \geq 0 \quad j = 1, \ldots, n\},$$

where η is an unbounded variable (representing the k-th largest share at the optimum) and s_j are additional nonnegative (deviational) variables.

If the sum of the $K_{inf} - 1$ largest shares is bounded by $1 - \varepsilon$, then there must be at least K_{inf} positive shares within the portfolio. Hence, any model can be extended to include a cardinality lower bound K_{inf} implemented by linear inequalities:

$$(K_{inf} - 1)\eta + \sum_{j=1}^{n} s_j \leq (1 - \varepsilon) \qquad (4.22a)$$

$$s_j \geq x_j - \eta, \ s_j \geq 0 \qquad\qquad j = 1, \ldots, n. \qquad (4.22b)$$

4.5 Logical or Decision Dependency Constraints

Decision dependency requirements (also defined *logical constraints*) impose relations among assets forcing their selection/exclusion according to specified preferences. These constraints are common in financial dealings. To be correctly modeled, they need the binary variables z_j already introduced for threshold and cardinality constraints. Usually they take one of the following forms:

- *Joint investment*: both assets i and j have to belong to the portfolio if asset l is selected

$$z_i + z_j \geq 2z_l. \qquad (4.23)$$

- *Mutually exclusive investment*: asset i cannot be selected if asset j is in the portfolio

$$z_i + z_j \leq 1. \qquad (4.24)$$

- *Contingent investment*: asset i can be selected only if asset j is in the portfolio

$$z_i \leq z_j. \qquad (4.25)$$

Combinations of these conditions are also possible, resulting in more complex relations. These investment restrictions can be an essential part of a diversification strategy and are often applied to investments in mutual funds.

4.6 Notes and References

Many papers on extensions of both Markowitz model and linear risk or safety based models were presented that considered cardinality constraints and thresholds. The cardinality constraints were introduced for Markowitz model in Bienstock (1996), Chang et al. (2000), Fieldsend et al. (2004), Jobst et al. (2001), Kumar et al. (2010), Lee and Mitchell (2000), Li et al. (2006), and Liu and Stefek (1995). All these models used shares to represent the investment. Linear models including cardinality constraints were proposed in Angelelli et al. (2008, 2012) and Speranza (1996) where amounts were used. Threshold constraints can be found in Chang et al. (2000), Kellerer et al. (2000), Mansini and Speranza (1999), and Young (1998). The idea of enforcing portfolio diversification by applying the CVaR concept to the right tail of the distribution of portfolio shares by using linear constraints was proposed in Mansini et al. (2007).

An overview of the papers presenting models with real features can be found in Mansini et al. (2014).

Chapter 5
Rebalancing and Index Tracking

5.1 Introduction

In the previous chapters we have analyzed the single-period portfolio optimization problem and provided LP or MILP models. In this chapter, our interest is focused on some related problems that can still be formulated as LP or MILP models.

A basic underlying assumption of the models already presented is that, at the time of the investment, the investor has an amount of money available but does not hold any portfolio that may influence his/her decision on how to use it. In this chapter, we consider the case where the investor already owns a portfolio of assets and, due to changed market conditions and possibly to the availability of additional capital, is interested in modifying it by selling/purchasing shares or amounts of some assets. This problem is known as the *portfolio rebalancing problem.*

When the market is performing well, an investor may be satisfied to optimally select and keep, also for a long time, the same portfolio of assets. When the market is performing poorly, this behavior becomes less sensible. Essentially, an investor is interested in rebalancing for two main reasons:

- The market trend is negative, and the investor believes the current portfolio will perform poorly in the future and its performance may be improved by modifying it;
- The market trend is positive, but the investor believes a modification of the current portfolio is appropriate because a better future performance may be achieved taking advantage of new information.

When rebalancing, an important aspect that cannot be ignored, at least in practice, is the role played by transaction costs. The rebalancing frequency is another feature that an investor has to take into account.

The performance of a market is captured by one or more *market indices*, where an index is composed by a subset of assets that represent the market. The models presented in the previous chapters aim at finding a trade-off between the return and the risk/safety of the portfolio, but in no way consider the market conditions explicitly. This is another important aspect that we have to analyze.

The aspiration of an investor may be to achieve a high expected return if the market conditions are very good, whereas in poor conditions a much lower expected return may be a sensible goal. This can be obtained replicating as closely as possible the performance of a market index. An investor may decide to invest money in institutional funds that fully replicate an index by investing in all of the assets that compose the index. Practical experience has made evident to many investors and financial institutions that achieving the same performance of an index may be a challenging task. The *index tracking problem* is the problem of selecting a set of assets that tries to match the return achieved by a benchmark index avoiding to invest in all of its assets. The goal is to mimic the index as closely as possible, while limiting the number of assets held in the portfolio and hence the associated transaction costs. When the goal is to exceed the performance of an index we talk about the *enhanced index tracking problem*. In this case, we may try to exceed the index performance, possibly by a specified excess return. In this chapter, we provide mathematical formulations for both index and enhanced index tracking problems, and discuss some important modelling issues.

At the end of the chapter, we briefly consider the case of long/short positions in portfolio holdings. In the previous chapters, we have assumed that amounts invested in assets are all non-negative. Considering a portfolio optimization problem that includes long/short positions means to assume that the sign of the investment in an asset is not constrained. A negative value is interpreted as a *short position*, whereas a positive value a *long position*.

5.2 Portfolio Rebalancing

We consider an investor who is interested in rebalancing a portfolio to take advantage of changed market conditions.

Essentially, with respect to the single period portfolio selection problem (*buy and hold strategy*), in the portfolio rebalancing problem, the investor makes investment decisions starting from possibly two different situations:

- Some additional capital is available for the rebalancing;
- No additional capital is available and consequently some assets must be sold, possibly partially, to permit investment in others (*self-financing portfolio balancing*).

From a risk/return efficient frontier point of view, the rebalancing decision may come from the observation that the currently held portfolio (which was located on the efficient frontier when originally built) now deviates from the present efficient frontier. Through rebalancing the investor invests/disinvests to move the portfolio towards the present efficient frontier. Although selling and buying shares or amounts

5.2 Portfolio Rebalancing

of assets usually implies transaction costs, it may be beneficial to rebalance a portfolio.

We indicate with $\kappa_j^0, j = 1, \ldots, n$, the portfolio available at the time of the rebalancing decision. In other words, the parameters $\kappa_j^0, j = 1, \ldots, n$, identify the portfolio composition before the rebalancing. A capital B, $B \geq 0$, is available to be used in the rebalancing. The situation in which $B = 0$ corresponds to the self-financing case.

The decision variables $\kappa_j, j = 1, \ldots, n$, represent, without loss of generality, the fractional number of units invested in each asset of the portfolio after the rebalancing. If an asset j is sold then $\kappa_j - \kappa_j^0 < 0$, whereas if it is purchased $\kappa_j - \kappa_j^0 > 0$. The absolute value $|\kappa_j - \kappa_j^0|$ is the amount of asset j sold or purchased, and thus indicates the change underwent by that asset. If no transaction occurs on asset j, obviously $\kappa_j = \kappa_j^0$. We may also want to limit the change for some asset $j, j = 1, \ldots, n$, to a predefined amount γ_j depending on the asset itself. In such a case the new portfolio will keep a positive position on that asset as follows:

$$\kappa_j^0 - \gamma_j \leq \kappa_j \leq \kappa_j^0 + \gamma_j.$$

More frequently, however, a limit is imposed only as an upper bound, so that an asset can be completely disinvested if convenient.

When rebalancing an existing portfolio one may assume that trading is free, that is buying/selling assets occur at zero transaction cost. Clearly, in real world, this assumption is unrealistic and the investor needs to account for transaction costs. The transaction costs measure the price the investor pays in order to move from the existing portfolio to the new one. As for the selection of a portfolio in a buy and hold strategy, also in this case the investor needs to account for the number of time periods he intends to hold the new portfolio.

In the following model we consider two types of cost. We first assume that any transaction, a sale or a purchase, on asset j incurs a fixed transaction cost $f_j, f_j \geq 0$. If $f_j = 0$, no fixed cost is charged. Moreover, a proportional cost $c_j, c_j \geq 0$, is applied to the transacted amount. Then, the proportional cost paid for asset j is calculated as

$$c_j q_j |\kappa_j - \kappa_j^0|,$$

where we recall that q_j is the quotation of asset j at the rebalancing time. To linearize such an expression, we introduce a non-negative variable, δ_j, to express the traded amount $|\kappa_j - \kappa_j^0| = \max\{\kappa_j - \kappa_j^0, \kappa_j^0 - \kappa_j\}$, for each asset $j = 1, \ldots, n$. This can be modelled by the two constraints:

$$\delta_j \geq (\kappa_j - \kappa_j^0)$$

$$\delta_j \geq -(\kappa_j - \kappa_j^0).$$

In the optimal solution, $\delta_j, j = 1, \ldots, n$, will take the value $|\kappa_j - \kappa_j^0|$, provided that the optimization model pushes the value of δ_j to its minimum feasible value.

To model the fixed cost $f_j, j = 1, \ldots, n$, we use the binary variable $z_j, j = 1, \ldots, n$, that takes value 1 if the investor buys or sells asset j, that is if $\delta_j > 0$, and 0 otherwise. As performance measure of the portfolio we adopt the CVaR. The model can be modified to accommodate a different performance measure. The parameter γ_j measures the maximum number of asset units that can be traded. The probabilities p_t and the variables d_t have the same meaning adopted in previous chapters. In the model we deduct the transaction costs from the expected return. As a consequence, such costs are deducted from the ending value of the portfolio.

The *rebalancing model* is as follows:

$$\max y - \frac{1}{\beta} \sum_{t=1}^{T} p_t d_t \tag{5.1a}$$

$$y - \sum_{j=1}^{n} r_{jt} q_j \kappa_j + \sum_{j=1}^{n} c_j q_j \delta_j + \sum_{j=1}^{n} f_j z_j \leq d_t \qquad t = 1, \ldots, T \tag{5.1b}$$

$$\sum_{j=1}^{n} r_j q_j \kappa_j - \sum_{j=1}^{n} c_j q_j \delta_j - \sum_{j=1}^{n} f_j z_j \geq \mu_0 \sum_{j=1}^{n} q_j \kappa_j \tag{5.1c}$$

$$\sum_{j=1}^{n} q_j \kappa_j = \sum_{j=1}^{n} q_j \kappa_j^0 + B \tag{5.1d}$$

$$\delta_j \geq (\kappa_j - \kappa_j^0) \qquad\qquad j = 1, \ldots, n \tag{5.1e}$$

$$\delta_j \geq -(\kappa_j - \kappa_j^0) \qquad\qquad j = 1, \ldots, n \tag{5.1f}$$

$$\delta_j \leq \gamma_j z_j \qquad\qquad j = 1, \ldots, n \tag{5.1g}$$

$$d_t \geq 0 \qquad\qquad t = 1, \ldots, T \tag{5.1h}$$

$$\kappa_j \geq 0 \qquad\qquad j = 1, \ldots, n \tag{5.1i}$$

$$\delta_j \geq 0 \qquad\qquad j = 1, \ldots, n \tag{5.1j}$$

$$z_j \in \{0, 1\} \qquad\qquad j = 1, \ldots, n. \tag{5.1k}$$

The objective function (5.1a), along with constraints (5.1b) and (5.1h), determines the maximization of the safety measure, where y is an auxiliary (unbounded) variable representing the β-quantile at the optimum and, as already mentioned, variables $d_t, t = 1, \ldots, T$, measure the deviation from the β-quantile. Constraint (5.1c) defines the net portfolio return and imposes that it has to be greater than or equal to the required expected return $\mu_0 \sum_{j=1}^{n} q_j \kappa_j$, where $\sum_{j=1}^{n} q_j \kappa_j$ is the amount invested in the rebalanced portfolio. The budget constraint (5.1d) requires that the capital invested has to be equal to the value of the initial portfolio at the rebalancing

5.2 Portfolio Rebalancing

time ($\sum_{j=1}^{n} q_j \kappa_j^0$) plus the additional capital B. Constraints (5.1e) and (5.1f) define each variable δ_j to be not less than the absolute difference $|\kappa_j - \kappa_j^0|$. Constraints (5.1g) impose an upper bound γ_j on the number of units of asset j that can be traded, $j = 1, \ldots, n$, and simultaneously defines the values of binary variables z_j, $j = 1, \ldots, n$. It forces variable z_j to take value 1 when $\delta_j \neq 0$, i.e. when the position of the asset j in the portfolio is changed ($|\kappa_j - \kappa_j^0| \neq 0$). If $z_j = 1$ the fixed cost f_j is included in the net mean return expression (constraint (5.1c)) and in the objective function through constraints (5.1b). When $\kappa_j = \kappa_j^0$, then constraints (5.1e) and (5.1f) let δ_j to take value 0 thereby allowing z_j to take any value in constraint (5.1g). In such case, z_j will take value 0. This is true since the objective function maximizes the mean of a specified size (quantile) of the worst portfolio realizations computed net of the transaction costs, and thus enforces not paying transaction costs whenever possible. This remains valid for any safety measure maximization, while in case a dispersion risk measure is minimized some more complicated modeling techniques might be necessary. Finally, constraints (5.1h)–(5.1k) are non-negative and binary conditions. It can be noticed that non-negativity requirements (5.1j) are redundant, as they follow directly from (5.1e) and (5.1f). Nevertheless, we keep them for the sake of completeness.

There are two important aspects that have to be further discussed and that concern transaction costs.

The first aspect is related to the way we deduct costs. In the described model, we deduct transaction costs from the expected return constraint. We have discussed alternative ways to introduce transaction costs in a portfolio model (see Chap. 3). Frequently, in a rebalancing problem, investors prefer to directly limit the amount of money spent for transaction costs to a fixed percentage ν ($0 < \nu < 1$) of the value of the existing portfolio plus the additional capital available B. In such a case costs have to be deducted from the return constraint and instead the following constraints have to be added to the model:

$$\sum_{j=1}^{n} c_j q_j \delta_j + \sum_{j=1}^{n} f_j z_j \leq \nu (\sum_{j=1}^{n} q_j \kappa_j^0 + B) \tag{5.2a}$$

$$\sum_{j=1}^{n} q_j \kappa_j = \sum_{j=1}^{n} q_j \kappa_j^0 + B - (\sum_{j=1}^{n} c_j q_j \delta_j + \sum_{j=1}^{n} f_j z_j). \tag{5.2b}$$

The first constraint limits the amount of the total transaction costs incurred, whereas the second one imposes that the value of the portfolio after trading is equal to the value before trading ($\sum_{j=1}^{n} q_j \kappa_j^0 + B$), minus the transaction costs incurred.

The second aspect to consider is related to the proportional transaction costs. In practice, more generally in our model, the proportional cost for the case of purchasing may be different from that of selling. Let us define $c_j^+, c_j^-, j = 1, \ldots, n$, these two proportional costs, respectively. In order to correctly model this *V-shaped cost* function we need to introduce two sets of variables $\delta_j^+, \delta_j^-, j = 1, \ldots, n$, to distinguish the amount invested or disinvested from asset j, respectively. Instead of

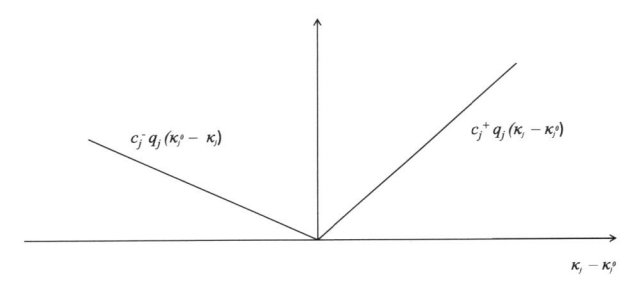

Fig. 5.1 V-shape cost function

constraints (5.1e) and (5.1f) we need to introduce the following ones:

$$\delta_j^+ \geq \kappa_j - \kappa_j^0, \quad \delta_j^+ \geq 0 \qquad j = 1, \ldots, n \qquad (5.3a)$$

$$\delta_j^- \geq \kappa_j^0 - \kappa_j, \quad \delta_j^- \geq 0 \qquad j = 1, \ldots, n, \qquad (5.3b)$$

and substitute the term $\sum_{j=1}^{n} c_j q_j \delta_j$ with $\sum_{j=1}^{n} c_j^+ q_j \delta_j^+ + \sum_{j=1}^{n} c_j^- q_j \delta_j^-$ in all the constraints. When $\kappa_j - \kappa_j^0 > 0$, then the maximization imposed by the safety measure will push δ_j^+ to take exactly the minimum value $\kappa_j - \kappa_j^0$, whereas δ_j^- will be forced to zero as the most convenient value. Similarly, for the case $\kappa_j^0 - \kappa_j > 0$ where $\delta_j^- > 0$ and δ_j^+ will be forced to zero. Figure 5.1 shows the classical V-shape structure of the proportional costs in a rebalancing problem.

Note that, given a well designed model considering a safety measure as objective function, the previous constraints (5.3a)–(5.3b) can be substituted with the equations

$$\kappa_j - \kappa_j^0 = \delta_j^+ - \delta_j^-, \quad \delta_j^+ \geq 0, \, \delta_j^- \geq 0 \quad j = 1, \ldots, n,$$

since complementarity conditions $\delta_j^+ \delta_j^- = 0, j = 1, \ldots, n$ are always satisfied at optimality.

Finally, one can observe that in many financial markets assets can be traded only in an integer number of units (see also the discussion on minimum transaction lots in Chap. 4). This would imply to transform continuous variables $\kappa_j, j = 1, \ldots, n$, into integer ones increasing the computational complexity of the problem.

5.3 Index Tracking

In modern financial stock exchanges, market indices have become standard benchmarks for evaluating the performance of a fund manager. Over the last few years, the number of funds managed by index-based investment strategies has increased tremendously in different economies. Traditionally, a fund manager who tries to replicate the behaviour of an index of a specific financial market is said to apply a *passive management strategy*, also referred to as index tracking. Models for index tracking aim at creating a portfolio mimicking the performance of a given index as closely as possible by minimizing a function (*the tracking error*) that measures how closely the portfolio tracks the market index.

Before presenting the models, let us briefly introduce the concept of market index and the way indices are constructed.

5.3.1 Market Index

A *stock index* or *stock market index*, or simply *market index*, is a measurement of the value of a section of the stock market. It is computed from the prices of selected stocks (typically a weighted average). It is a tool used by investors and financial managers to describe the market, and to compare the return on specific investments.

Market indices may be classified in many ways. A world or global market index includes typically large companies without any regard for where they are domiciled or traded. Two examples are MSCI World and S&P Global 100. A national index represents the performance of the financial market of a given nation and reflects the investor sentiment on the state of its economy. The most regularly quoted market indices are national indices composed of the assets of large companies listed on the largest stock exchanges of a nation, such as the American S&P100 and S&P500, the Japanese Nikkei225, the British FTSE100, the Chinese HangSeng31, the German DAX100 and the American Russell2000 and Russell3000. As the name of the index indicates the number of assets included in each index ranges from 31, composing the Hang Seng index, to 2151, composing the Russell3000 index.

Other indices may be regional, such as the FTSE Developed Europe Index or the FTSE Developed Asia Pacific Index.

The concept of index may be extended well beyond a stock exchange. The Wilshire 5000 Index, the original total market index, represents the stocks of nearly every publicly traded company in the United States (US), including US stocks traded on the New York Stock Exchange, NASDAQ and American Stock Exchange.

More specialized indices exist measuring the performance of specific sectors of the market. Some examples include the Wilshire US REIT which tracks more than 80 American real estate investment trusts, and the Morgan Stanley Biotech Index which consists of 36 American firms in the biotechnology industry. Other indices may be associated with companies of a certain size, a certain type of management, or even more specialized criteria.

5.3.2 An Index Tracking Model

The most intuitive way to track an index is through perfect replication, i.e. by building a portfolio composed of the same assets that constitute the index and in the same proportions. However, this strategy is too difficult to implement in practice as market indices usually include large numbers of assets. Furthermore, the perfect replication implies that when the composition of the index is revised, the portfolio should be rebalanced accordingly. Perfect replication of an index would imply very high transaction costs. Thus, the best strategy an investor may adopt is that of tracking an index, holding fewer assets than those included in the index. This is the approach we follow in constructing an index tracking model.

We indicate the value (quotation) of asset j in scenario t, $t = 1, \ldots, T$, as q_{jt} and the value (price) of the index we are tracking as I_t. As for portfolio optimization problem, we use an historic look-back approach: we select the portfolio, i.e. the assets and the quantity to invest in each of them to best track the index over the last T historical periods, and then we will hold the selected portfolio in the immediate future. Scenarios are assumed to be all equally probable, i.e. the probability of each scenario t is computed as $p_t = 1/T$, $t = 1, \ldots, T$.

The decision variables κ_j, $j = 1, \ldots, n$, represent the fractional number of asset units selected in the portfolio. We assume that short sales are not allowed, i.e. κ_j, $j = 1, \ldots, n$, are constrained to be non-negative. We are interested in selecting the optimal portfolio composition that minimizes the in-sample tracking error. Let $\bar{C}$ be the capital available. If asset j is selected in the optimal portfolio, then the amount invested must be within a lower bound, denoted as L_j, and an upper bound, denoted as U_j. To this aim, we introduce a binary variable z_j, $j = 1, \ldots, n$, that takes value 1 if the investor holds any fraction of asset j, and 0 otherwise.

Our approach focuses on the normalized trajectory I_t/I_T, $t = 1, \ldots, T$, of index values. Scenario T corresponds to the date the optimal portfolio is chosen. The former trajectory is then compared with the corresponding trajectory of portfolio values $W_t = \sum_{j=1}^{n} q_{jt}\kappa_j$, $t = 1, \ldots, T$, to compute the tracking error. Specifically, we adopt the MAD to measure the tracking error between the index and the new portfolio expressed in terms of amounts

$$TrE(\kappa) = \sum_{t=1}^{T} p_t |\theta I_t - \sum_{j=1}^{n} q_{jt}\kappa_j|,$$

where the parameter $\theta = \bar{C}/I_T$ is used to scale the value of the index to be tracked. Intuitively, the parameter θ represents the amount the investor can buy in scenario T of a fictitious asset whose price is equal to I_T.

The index tracking model aims at minimizing the non-linear function $TrE(\kappa)$. The function can be linearized introducing two non-negative variables d_t^- and d_t^+, for each scenario t, $t = 1, \ldots, T$. Variable d_t^- takes positive value when $\theta I_t > \sum_{j=1}^{n} q_{jt}\kappa_j$ and measures the downside deviation of the portfolio from the

5.3 Index Tracking

market index value under scenario t, and takes value zero otherwise, i.e. $d_t^- = \max\{\theta I_t - \sum_{j=1}^n q_{jt}\kappa_j, 0\}$. Variable d_t^+ takes positive value when $\sum_{j=1}^n q_{jt}\kappa_j > \theta I_t$ and measures the upside deviation of the portfolio from the market index in scenario t, and takes value zero otherwise, i.e. $d_t^+ = \max\{\sum_{j=1}^n q_{jt}\kappa_j - \theta I_t, 0\}$. To work correctly, variables d_t^- and d_t^+ have to satisfy the following conditions:

$$d_t^- - d_t^+ = \theta I_t - \sum_{j=1}^n q_{jt}\kappa_j, \quad d_t^-, d_t^+ \geq 0 \qquad t = 1, \ldots, T \qquad (5.4)$$

$$d_t^- d_t^+ = 0 \qquad t = 1, \ldots, T, \qquad (5.5)$$

that lead to the following representation of the mean absolute deviation:

$$d_t^- + d_t^+ = \left| \theta I_t - \sum_{j=1}^n q_{jt}\kappa_j \right| \quad t = 1, \ldots, T.$$

Actually, it can be seen that complementarity conditions (5.5) are not necessary, as being enforced by the minimization of the mean absolute deviation $\sum_{t=1}^T (d_t^- + d_t^+)$ that will be set as objective function.

The index tracking model can then be formulated as follows:

$$\min \sum_{t=1}^T p_t (d_t^- + d_t^+) \qquad (5.6a)$$

$$d_t^- - d_t^+ = \theta I_t - \sum_{j=1}^n q_{jt}\kappa_j \qquad t = 1, \ldots, T \qquad (5.6b)$$

$$\sum_{j=1}^n q_{jT}\kappa_j = \bar{C} \qquad (5.6c)$$

$$\sum_{j=1}^n z_j \leq K \qquad (5.6d)$$

$$L_j z_j \leq \kappa_j q_{jT} \leq U_j z_j \qquad j = 1, \ldots, n \qquad (5.6e)$$

$$\kappa_j \geq 0 \qquad j = 1, \ldots, n \qquad (5.6f)$$

$$d_t^- \geq 0 \qquad t = 1, \ldots, T \qquad (5.6g)$$

$$d_t^+ \geq 0 \qquad t = 1, \ldots, T \qquad (5.6h)$$

$$z_j \in \{0, 1\} \qquad j = 1, \ldots, n. \qquad (5.6i)$$

The objective function (5.6a) minimizes the sum of the upper and lower mean deviations of the portfolio from the market index as defined in constraints (5.6b),

(5.6g)–(5.6h). Constraint (5.6c) is the budget constraint imposing that the capital invested is equal to the capital available and that the portfolio purchase is done at time T. In order to control the number of assets selected in the portfolio, constraint (5.6d) imposes that such a number has to be lower than or equal to a predefined value K. Constraints (5.6e) ensure that if asset j is selected, i.e. $\kappa_j > 0$, then the associated binary variable z_j takes value 1. If this is the case, the amount invested in asset j in the optimal portfolio is constrained to be within the defined lower and upper bounds. Conversely, if asset j is not selected in the portfolio, i.e. $\kappa_j = 0$, then variable z_j takes value equal to 0. Finally, the remaining constraints are non-negative and binary conditions on variables.

Note that one may consider to minimize only the lower deviation from the index instead of the absolute deviation, by simply considering variables d_t^-, $t = 1, \ldots, T$. While it is obvious that the new model has some advantages, there is, however, no guarantee on how well this model can track the index since it ignores the upside deviation.

We may also track an index rate of return. In this case, we can reformulate the index tracking model by considering the mean absolute deviation of the portfolio rate of return from the index rate of return. If we ignore transaction costs, we can simply refer to a portfolio in terms of shares $x_j, j = 1, \ldots, n$. More precisely, we denote the r.v. representing the rate of return of the market index as R^I, its realization under scenario t as r_t^I, with $t = 1, \ldots, T$, and its mean rate of return as $\mu^I = \sum_{t=1}^{T} r_t^I p_t$. Then, the tracking error can be computed as the sum of the deviations of the portfolio rate of return from the rate of return of the index under the different scenarios

$$\min \sum_{t=1}^{T} p_t(d_t^+ + d_t^-) \tag{5.7a}$$

$$d_t^- \geq r_t^I - \sum_{j=1}^{n} r_{jt}x_j, \ d_t^- \geq 0 \qquad t = 1, \ldots, T \tag{5.7b}$$

$$d_t^+ \geq -(r_t^I - \sum_{j=1}^{n} r_{jt}x_j), \ d_t^+ \geq 0 \qquad t = 1, \ldots, T, \tag{5.7c}$$

or from a target represented by the index mean rate of return (substituting in constraints (5.7b)–(5.7c) r_t^I with μ^I).

We can now consider the extension of the previous index tracking model (5.6a)–(5.6i) to include transaction costs. Let us assume that if the investor buys κ_j fractional units of asset j, a proportional transaction cost $c_j q_{jT} \kappa_j$ is incurred, where c_j is the unitary transaction cost paid for purchasing asset j. Additionally, a fixed transaction cost, denoted as f_j, $j = 1, \ldots, n$, is paid by the investor if asset j is traded. To model the fixed cost, we use the above mentioned binary variable z_j, $j = 1, \ldots, n$. Variable z_j takes value 1 if the investor buys asset j, and 0 otherwise.

5.4 Enhanced Index Tracking

We have seen in Chap. 3 that there are different ways to include the transaction costs in a model. Since no constraint on portfolio expected return is necessary in the index tracking model, we constrain the total transaction costs paid by the investor to be less than or equal to a proportion ϑ ($0 < \vartheta < 1$) of the capital available by introducing the following constraint:

$$\sum_{j=1}^{n} c_j q_{jT} \kappa_j + \sum_{j=1}^{n} f_j z_j \leq \vartheta \bar{C}. \tag{5.8}$$

Our choice here is also motivated by the advantage of clearly separating the capital invested in the portfolio from that spent in transaction costs. This is consistent with some fund management applications where transaction costs are paid out of a separate account. Therefore, the right hand side in constraint (5.8) has to be intended as the amount in the separate account the decision maker is prepared to pay in transaction costs.

Finally, in the described model, without loss of generality, we have considered asset variables representing a fractional number of units for each asset. We may substitute them with integer variables measuring the number of units bought for each asset.

In the presented model we construct ex-novo a portfolio that tracks the market index, starting with a capital available $\bar{C}$ only. The problem can be generalized to the rebalancing case assuming to start with an existing portfolio and possibly additional capital. In this case the index tracking problem becomes that of updating (buying or selling) the current portfolio composition to best track the market index.

5.4 Enhanced Index Tracking

Enhanced indexing is an expression that describes strategies adopted to outperform an index. Enhanced indexing attempts to generate modest excess returns with respect to an index. Usually, a fund manager who makes specific investments with the goal of outperforming a benchmark index is said to implement an *active management* strategy.

The enhanced index tracking problem aims at selecting a portfolio outperforming the value of a given market index. The intrinsic nature of the enhanced index tracking problem is bi-objective: to maximize the excess return of the portfolio over the benchmark, on one hand, and to minimize the tracking error, on the other hand. Thus, the objective is expressed as the maximization of the over-performance with respect to the market index, whereas the tracking error is bounded in the constraints to be below a chosen threshold.

A straightforward way to account for enhanced indexation is to simply select a portfolio that tracks an *index-plus-alpha* portfolio. The term index-plus-alpha portfolio refers to a portfolio that outperforms the benchmark by a given, typically small, amount α. In the following model α is a free variable to be maximized, while the tracking error is measured as downside deviation. In a similar way, other risk measures described in this book can be used as well.

A model for enhanced index tracking problem can be formulated as:

$$\max \alpha \tag{5.9a}$$

$$\sum_{t=1}^{T} p_t d_t^- \leq \xi \tag{5.9b}$$

$$d_t^- \geq \theta(1+\alpha)I_t - \sum_{j=1}^{n} q_{jt}\kappa_j \qquad t = 1, \ldots, T \tag{5.9c}$$

$$\sum_{j=1}^{n} q_{jT}\kappa_j = \bar{C} \tag{5.9d}$$

$$\sum_{j=1}^{n} z_j \leq K \tag{5.9e}$$

$$L_j z_j \leq \kappa_j q_{jT} \leq U_j z_j \qquad j = 1, \ldots, n \tag{5.9f}$$

$$\kappa_j \geq 0 \qquad j = 1, \ldots, n \tag{5.9g}$$

$$d_t^- \geq 0 \qquad t = 1, \ldots, T \tag{5.9h}$$

$$z_j \in \{0, 1\} \qquad j = 1, \ldots, n. \tag{5.9i}$$

The quantity $(1 + \alpha)I_t$ represents the value of the index-plus-alpha portfolio in scenario $t, t = 1, \ldots, T$. The objective function (5.9a) maximizes the free variable α representing the excess performance with respect to the market index in each scenario $t = 1, \ldots, T$. Parameter ξ, used to bound the tracking error, indicates the acceptable level of downside deviation from the index and is chosen by the decision maker. As for the index tracking we may also consider the rebalancing variant of the problem. It is obvious how, at least in practice, in an attempt to beat the benchmark a frequent trading may be necessary. In such a case transaction costs cannot be ignored.

Finally, we may consider returns instead of absolute values. Actually, the investor may be interested in determining a portfolio that outperforms the rate of return of the market index by a given excess return $\tilde{\alpha}$. To model this goal, rather than the market index rate of return R^I, one can use some reference r.v. $R^{\tilde{\alpha}} = R^I + \tilde{\alpha}$ representing the rate of return beating the market index return by $\tilde{\alpha}$, with realization $r_t^{\tilde{\alpha}} = r_t^I + \tilde{\alpha}$ under scenario t, with $t = 1, \ldots, T$, and mean rate of return $\mu^{\tilde{\alpha}} = \sum_{t=1}^{T} r_t^{\tilde{\alpha}} p_t$.

5.5 Long/Short Portfolios

In all the presented models we have assumed that shares or amounts invested in assets are non-negative (long positions).

The idea of using short positions in conjunction with long positions in a portfolio optimization framework, aims at identifying an optimized portfolio in which long and short positions are determined jointly within the optimization model. The target is to obtain a portfolio with possibly better performance compared to the case of the purchase only portfolio. We try to clarify the benefit we can get from holding a short position in a portfolio with an example.

An investor has an extremely negative view on the future performance of a given asset. The quotation of this asset is believed to be overestimated and to deteriorate in the future. If only a portfolio with long positions can be held, then the most an investor can do to benefit from this insight is to exclude the asset from the portfolio. If, on the contrary, the portfolio allows short positions, then the best is to borrow shares on this asset with the expectation to pay them back at a lower price given the poor expectations on the asset.

To represent a long/short portfolio with n assets one should define n free variables x_j, each representing the share of capital (either positive or negative) to be invested into asset j. When the investor purchases asset j, then $x_j > 0$ whereas when he/she sells the asset short, then $x_j < 0$. Technically, it can be easily implemented by introducing two auxiliary variables, for each j, such that

$$x_j = x_j^+ - x_j^-, \ x_j^+ \geq 0, \ x_j^- \geq 0.$$

The introduction of short positions requires some additional constraints, some of them requested by the market regulator, with respect to the classic portfolio optimization as:

- To take short the investor has to pay a deposit, usually proportional to the amount of short investment, which affects the budget constraint;
- The sum of the long positions plus the sum of the (absolute value of) short positions must not exceed a predefined value;
- If an asset is difficult to borrow it is better to limit the amount of its short position or even not to permit short positions in that particular asset;
- Portfolio return has to be computed summing the return of the assets hold in long position minus the return of assets sold short;
- If it is not allowed to take both short and long positions on the same asset, then such a portfolio is referred as trimmed and the complementarity conditions $x_j^+ x_j^- = 0$ have to be added; if it is allowed to hold both positions for each asset, then the portfolio is called an untrimmed portfolio.

5.6 Notes and References

The importance of intermediate adjustment opportunities was already recognized in Markowitz (1952). He suggested that finding the optimal multi-period management strategy is, in principle, a dynamic programming problem. Smith (1967) introduced a model of portfolio revision with transaction costs. His approach aimed at modifying Markowitz one-period problem and then applying the solution period by period. Chen et al. (1971) pointed out that this approach fails to take into account the effect that future adjustment opportunities should have on current investment decisions. They introduced a dynamic programming approach and succeeded in formulating the problem with proportional transaction costs. Pogue (1970) first dealt with the proportional brokerage fees involved in revising an existing portfolio.

For portfolio rebalancing models with transaction costs in a mean-variance framework we refer to Adcock and Meade (1994) and Woodside-Oriakhi et al. (2013). Very few contributions can be found on models based on LP computable risk measures for portfolio rebalancing, see Konno and Wijayanayake (2001) for MAD-based models and Guastaroba et al. (2009a) for a CVaR-based model.

Linear programming models for index tracking and enhanced index tracking models were presented in Guastaroba and Speranza (2012), and more recently, in Guastaroba et al. (2014), where Omega ratio was proposed as performance measure for enhanced index tracking. An alternative view on index tracking that minimized ordinary least-squares regression was introduced by Canakgoz and Beasley (2009). For enhanced indexation using modified Sortino ratio we refer to Meade and Beasley (2011). In Koshizuka et al. (2009) the index-plus-alpha portfolio was described together with its use. The problem of selecting a so-called absolute portfolio (i.e., market neutral portfolio) is that of finding a portfolio of assets that yields a good return irrespective of how the underlying market index performs. We refer to Valle et al. (2014a,b) for the introduction of the concept of absolute return portfolio and its use in enhanced indexation.

A long-short portfolio optimization problem under non-convex transaction cost using MAD as risk function was introduced by Konno et al. (2005).

Chapter 6
Theoretical Framework

6.1 Introduction

The problem of choice among investment opportunities or portfolios having uncertain returns represents a special problem of comparison of uncertain outcomes. The latter is one of fundamental interest in decision theory. Assuming basic rationality axioms, allowing an investor to order all conceivable risky prospects in a continuous and consistent way, actually implies that the investor behavior can be modeled as driven by a real-valued utility function used to compare outcomes. This is the so-called *expected utility* hypothesis. In practice, however, it is almost impossible to find the actual utility function explicitly. Nevertheless, one may identify criteria or properties of the utility function defined by preferences shared by all rational investors. For example, the utility function can be assumed to be non-decreasing, which means that more is preferred to less (maximization). Such general preferences define a partial order of risky outcomes, the so-called stochastic dominance. The stochastic dominance approach provides a theoretical framework but not a simple computational recipe to search for the best portfolio. The portfolio optimization models considered in this book are based on the mean-risk or mean-safety approach quantifying the problem in the form of only two criteria: the mean, representing the expected outcome, and the risk or the safety, a scalar measure of the variability of outcomes. It is important to understand whether the models satisfy the partial order of risky outcomes. When they do, they are said to be consistent with the stochastic dominance. Unfortunately, for typical dispersion statistics as risk measures, like variance, the mean-risk approach consistency cannot be proved. A different, though related, concept is that of coherent risk measure and is defined through properties of a risk measure that capture the preferences of a rational investor.

It is out of the scope of this book to present these theories in depth. Nevertheless, we will introduce the basic concepts and discuss their relevance for the portfolio optimization models. In this chapter, we review the LP portfolio optimization models with respect to their consistency with the stochastic dominance rules and other axiomatic theories. In the next section, the risk averse preferences and the stochastic dominance relations are formalized. Further the consistency of the LP computable risk measures with the stochastic dominance is examined, and it is shown that the models based on a safety measure maximization are consistent. Finally, we examine LP computable measures with respect to the axiomatic model of the coherent risk measurement. Again, we show that the models based on the safety measure maximization meet the coherent measurement rules.

6.2 Risk Averse Preferences and Stochastic Dominance

The problem of optimal portfolio selection is related to the problem of comparing prospects described by random variables with known distributions. Basic axioms of the investor rationality, namely completeness, transitivity, independence and continuity, imply the possibility to define a utility function. This means that we can say that, for a rational investor, there exists a utility function u such that $R_{\mathbf{x}'}$ is preferred to $R_{\mathbf{x}''}$ if and only if $\mathbb{E}\{u(R_{\mathbf{x}'})\} \geq \mathbb{E}\{u(R_{\mathbf{x}''})\}$. Whereas it is impossible to find the utility function of a specific investor, it is important to identify criteria that are shared by all rational investors. Stochastic Dominance (SD) is a family of partial orders closely connected to the expected utility theory but they eliminate the need to explicitly specify a utility function. With stochastic dominance, random variables are ordered by point-wise comparison of functions constructed from their distribution functions.

The so-called *First-order Stochastic Dominance* (FSD) relation depends on the comparison of underachievement probabilities for all possible real targets τ:

$$R_{\mathbf{x}'} \succeq_{FSD} R_{\mathbf{x}''} \quad \Leftrightarrow \quad \mathbb{P}\{R_{\mathbf{x}'} \leq \tau\} \leq \mathbb{P}\{R_{\mathbf{x}''} \leq \tau\} \quad \text{for all } \tau.$$

Thus, in FSD uncertain returns are compared by point-wise comparison of the right-continuous cumulative distribution functions (cdf), $F_{\mathbf{x}}(\tau) = \mathbb{P}\{R_{\mathbf{x}} \leq \tau\}$, as follows:

$$R_{\mathbf{x}'} \succeq_{FSD} R_{\mathbf{x}''} \quad \Leftrightarrow \quad F_{\mathbf{x}'}(\tau) \leq F_{\mathbf{x}''}(\tau) \quad \text{for all } \tau.$$

We say that portfolio $\mathbf{x}'$ dominates $\mathbf{x}''$ under the FSD (and we write $R_{\mathbf{x}'} \succ_{FSD} R_{\mathbf{x}''}$) if $R_{\mathbf{x}'} \succeq_{FSD} R_{\mathbf{x}''}$ and $R_{\mathbf{x}''} \not\succeq_{FSD} R_{\mathbf{x}'}$. That means $R_{\mathbf{x}'} \succ_{FSD} R_{\mathbf{x}''}$ if $F_{\mathbf{x}'}(\tau) \leq F_{\mathbf{x}''}(\tau)$ for all τ, and there exists at least one τ_0 such that $F_{\mathbf{x}'}(\tau_0) < F_{\mathbf{x}''}(\tau_0)$. A feasible portfolio $\mathbf{x}^0 \in Q$ is called *FSD efficient* if there is no $\mathbf{x} \in Q$ such that $R_{\mathbf{x}} \succ_{FSD} R_{\mathbf{x}^0}$.

Alternatively, FSD can be characterized by point-wise comparison of quantiles. As the quantile function (see (2.15))

$$F_{\mathbf{x}}^{(-1)}(\beta) = q_\beta(\mathbf{x}) = \inf\{\tau : F_{\mathbf{x}}(\tau) \geq \beta\} \quad 0 < \beta \leq 1 \tag{6.1}$$

6.2 Risk Averse Preferences and Stochastic Dominance

is the left-continuous inverse of the cdf $F_\mathbf{x}$ we obtain

$$R_{\mathbf{x}'} \succeq_{FSD} R_{\mathbf{x}''} \quad \Leftrightarrow \quad F_{\mathbf{x}'}^{(-1)}(\beta) \geq F_{\mathbf{x}''}^{(-1)}(\beta) \quad \text{for all } 0 < \beta \leq 1.$$

The FSD relation represents general preferences that more is preferred to less. In terms of the expected utility, it defines the corresponding inequality for the entire class of non-decreasing utility functions:

$$R_{\mathbf{x}'} \succeq_{FSD} R_{\mathbf{x}''} \quad \Leftrightarrow \quad \mathbb{E}\{u(R_{\mathbf{x}'})\} \geq \mathbb{E}\{u(R_{\mathbf{x}''})\} \quad \text{for all non-decreasing functions } u.$$

Actually, if portfolio $\mathbf{x}'$ dominates $\mathbf{x}''$ under the FSD ($R_{\mathbf{x}'} \succ_{FSD} R_{\mathbf{x}''}$), then for any strictly increasing utility function u one gets $\mathbb{E}\{u(R_{\mathbf{x}'})\} > \mathbb{E}\{u(R_{\mathbf{x}''})\}$. Additionally, if neither $\mathbf{x}'$ nor $\mathbf{x}''$ FSD dominates the other, then there exist strictly increasing u and v, such that $\mathbb{E}\{u(R_{\mathbf{x}'})\} > \mathbb{E}\{u(R_{\mathbf{x}''})\}$ and $\mathbb{E}\{v(R_{\mathbf{x}'})\} < \mathbb{E}\{v(R_{\mathbf{x}''})\}$. That means, all rational investors will strictly prefer portfolio $\mathbf{x}'$ to $\mathbf{x}''$ if and only if portfolio $\mathbf{x}'$ FSD dominates $\mathbf{x}''$.

The FSD, that aims at characterizing investor preferences, does not take into account the attitude to risk. Investors are usually assumed to be risk averse, in the sense that a certain yield of value $\mathbb{E}\{R_\mathbf{x}\}$ is preferred to the risky return $R_\mathbf{x}$ for any distribution of $R_\mathbf{x}$. Such general risk averse preferences define the *Second-order Stochastic Dominance* (SSD) based on the point-wise comparison of the second order cumulative distribution functions derived from the cdf as $F_\mathbf{x}^{(2)}(\tau) = \int_{-\infty}^{\tau} F_\mathbf{x}(\xi) \, d\xi$ for real numbers τ:

$$R_{\mathbf{x}'} \succ_{SSD} R_{\mathbf{x}''} \quad \Leftrightarrow \quad F_{\mathbf{x}'}^{(2)}(\tau) \leq F_{\mathbf{x}''}^{(2)}(\tau) \quad \text{for all } \tau.$$

We say that portfolio $\mathbf{x}'$ dominates $\mathbf{x}''$ under the SSD (and we write $R_{\mathbf{x}'} \succ_{SSD} R_{\mathbf{x}''}$), if $R_{\mathbf{x}'} \succeq_{SSD} R_{\mathbf{x}''}$ and $R_{\mathbf{x}''} \not\succeq_{SSD} R_{\mathbf{x}'}$. A feasible portfolio $\mathbf{x}^0 \in Q$ is called *SSD efficient* if there is no $\mathbf{x} \in Q$ such that $R_\mathbf{x} \succ_{SSD} R_{\mathbf{x}^0}$. Thus, portfolio $\mathbf{x}'$ dominates $\mathbf{x}''$ under the SSD ($R_{\mathbf{x}'} \succ_{SSD} R_{\mathbf{x}''}$), if $F_{\mathbf{x}'}^{(2)}(\tau) \leq F_{\mathbf{x}''}^{(2)}(\tau)$ for all τ, with at least one strict inequality.

Values of the second order cumulative distribution function represent mean shortfalls to the corresponding targets: $F_\mathbf{x}^{(2)}(\tau) = \mathbb{E}\{\max\{\tau - R_\mathbf{x}, 0\}\}$. Therefore, the SSD relation may be expressed as the comparison of mean shortfalls to all possible targets:

$$R_{\mathbf{x}'} \succeq_{SSD} R_{\mathbf{x}''} \quad \Leftrightarrow \quad \mathbb{E}\{\max\{\tau - R_{\mathbf{x}'}, 0\}\} \leq \mathbb{E}\{\max\{\tau - R_{\mathbf{x}''}, 0\}\} \quad \text{for all } \tau.$$

Note that the latter corresponds to a point-wise comparison of all mean below-target deviations $\bar{\delta}_\tau(\mathbf{x})$ (2.49).

An alternative characterization of the SSD relation can be achieved with the so-called Absolute Lorenz Curves (ALC) which represent the second order quantile functions defined as

$$F_{\mathbf{x}}^{(-2)}(\beta) = \int_0^\beta F_{\mathbf{x}}^{(-1)}(\alpha)d\alpha \quad \text{for } 0 < \beta \leq 1 \quad \text{and} \quad F_{\mathbf{x}}^{(-2)}(0) = 0, \qquad (6.2)$$

where $F_{\mathbf{x}}^{(-1)}(\alpha)$ is the quantile function (6.1). The point-wise comparison of ALCs is equivalent to the SSD relation in the sense that

$$R_{\mathbf{x}'} \succeq_{SSD} R_{\mathbf{x}''} \quad \Leftrightarrow \quad F_{\mathbf{x}'}^{(-2)}(\beta) \geq F_{\mathbf{x}''}^{(-2)}(\beta) \quad \text{for all } 0 < \beta \leq 1.$$

The latter makes the SSD relation related not only to the expected utility model but also to the so-called dual theory of choice under risk or rank dependent utility.

The values of function $F_{\mathbf{x}}^{(-2)}(\beta)$ cumulate the values within the corresponding quantile tails, hence representing denormalized values of the corresponding CVaR measure (2.19):

$$F_{\mathbf{x}}^{(-2)}(\beta) = \beta M_\beta(\mathbf{x}) \quad \text{and} \quad M_\beta(\mathbf{x}) = \frac{1}{\beta} F_{\mathbf{x}}^{(-2)}(\beta).$$

The latter makes the CVaR directly related to the dual theory of choice under risk.

In terms of the expected utility, the SSD defines the corresponding inequality for the entire class of non-decreasing concave utility functions:

$$R_{\mathbf{x}'} \succeq_{SSD} R_{\mathbf{x}''} \quad \Leftrightarrow \quad \mathbb{E}\{u(R_{\mathbf{x}'})\} \geq \mathbb{E}\{u(R_{\mathbf{x}''})\} \quad \text{for all non-decreasing concave } u.$$

This limits the class of non-decreasing utility functions to concave functions thus satisfying inequality $u(\mathbb{E}\{R_{\mathbf{x}}\}) \geq \mathbb{E}\{u(R_{\mathbf{x}})\}$ for all $R_{\mathbf{x}}$ required to model risk aversion. If portfolio $\mathbf{x}'$ dominates $\mathbf{x}''$ under the SSD ($R_{\mathbf{x}'} \succ_{SSD} R_{\mathbf{x}''}$), then for any strictly increasing and strictly concave utility function u one gets $\mathbb{E}\{u(R_{\mathbf{x}'})\} > \mathbb{E}\{u(R_{\mathbf{x}''})\}$. Additionally, if neither $\mathbf{x}'$ nor $\mathbf{x}''$ SSD dominates the other, then there exist strictly increasing strictly concave u and v, such that $\mathbb{E}\{u(R_{\mathbf{x}'})\} > \mathbb{E}\{u(R_{\mathbf{x}''})\}$ and $\mathbb{E}\{v(R_{\mathbf{x}'})\} < \mathbb{E}\{v(R_{\mathbf{x}''})\}$. That means, all rational risk averse investors will strictly prefer portfolio $\mathbf{x}'$ to $\mathbf{x}''$ if and only if portfolio $\mathbf{x}'$ SSD dominates $\mathbf{x}''$.

Note that FSD implies SSD and inequality on the expected values:

$$R_{\mathbf{x}'} \succeq_{FSD} R_{\mathbf{x}''} \quad \Rightarrow \quad R_{\mathbf{x}'} \succeq_{SSD} R_{\mathbf{x}''} \quad \Rightarrow \quad \mathbb{E}\{R_{\mathbf{x}'}\} \geq \mathbb{E}\{R_{\mathbf{x}''}\}. \qquad (6.3)$$

The implications follow directly from the SD definitions, although they can be lucidly justified through the expected utility interpretations. $R_{\mathbf{x}'} \succeq_{FSD} R_{\mathbf{x}''}$ means that $\mathbb{E}\{u(R_{\mathbf{x}'})\} \geq \mathbb{E}\{u(R_{\mathbf{x}''})\}$ for all non-decreasing utility functions and thereby for all non-decreasing concave function either, which means the SSD relation. Moreover, it remains also valid for identity function u thus expressing the expectations inequality. Therefore, neither under FSD nor SSD a portfolio with higher mean return can be dominated by a portfolio with lower mean return.

6.3 Stochastic Dominance Consistency

In portfolio optimization, the comparison of random variables representing returns is related to the problem of choice among risky alternatives in a given feasible set Q. For instance, in the simplest problem the feasible set of random variables is defined as all convex combinations (weighted averages with non-negative weights totaling 1) of a given number of investment opportunities (assets). If $R_{\mathbf{x}'} \succ_{SSD} R_{\mathbf{x}''}$, then $R_{\mathbf{x}'}$ is preferred to $R_{\mathbf{x}''}$ within all risk-averse preference models where larger outcomes are preferred. It is therefore a matter of primary importance that a model for portfolio optimization be consistent with the SSD relation. Note that due to (6.3) the SSD consistency implies also the FSD consistency, related to all models where larger outcomes are preferred. If the risk measure $\varrho(\mathbf{x})$ is SSD consistent in the sense that

$$R_{\mathbf{x}'} \succeq_{SSD} R_{\mathbf{x}''} \quad \Rightarrow \quad \varrho(\mathbf{x}') \leq \varrho(\mathbf{x}''), \tag{6.4}$$

then except for portfolios with identical values of $\mu(\mathbf{x})$ and $\varrho(\mathbf{x})$, every efficient solution of the bicriteria mean-risk problem (1.3):

$$\max\{[\mu(\mathbf{x}), -\varrho(\mathbf{x})] : \mathbf{x} \in Q\} \tag{6.5}$$

is SSD efficient. The same applies then to the optimal solution of the corresponding ratio optimization problem (6.6):

$$\max \left\{ \frac{\mu(\mathbf{x}) - r_0}{\varrho(\mathbf{x})} : \mathbf{x} \in Q \right\}. \tag{6.6}$$

That means, for an SSD consistent risk measure, except for portfolios with identical values of $\mu(\mathbf{x})$ and $\varrho(\mathbf{x})$, the tangency portfolio is SSD efficient.

The SSD consistency conditions fulfill only very specific shortfall risk measures like lower partial moments and particularly the mean below-target deviation (2.49) that represents a single value of the second order cumulative distribution function $F^{(2)}$. This also justifies SSD consistency of the Omega maximization model (2.52), due to its equivalence to the standard ratio (tangent portfolio) model (6.6) for the mean below-target deviation measure with target τ replacing the risk-free rate of return r_0.

Typical dispersion type risk measures when minimized may lead to choices inconsistent even with the FSD and, thus, not acceptable for any preference related to maximization of returns (more better than less). Recall Example 1.2 showing two portfolios $\mathbf{x}'$ and $\mathbf{x}''$ (with rates of return given in percentage):

$$\mathbb{P}\{R_{\mathbf{x}'} = \xi\} = \begin{cases} 1, & \xi = 1\% \\ 0, & \text{otherwise} \end{cases} \quad \mathbb{P}\{R_{\mathbf{x}''} = \xi\} = \begin{cases} 1/2, & \xi = 3\% \\ 1/2, & \xi = 5\% \\ 0, & \text{otherwise.} \end{cases}$$

Although the two portfolios are both mean-risk efficient for any dispersion type measure, the risk-free portfolio $\mathbf{x}'$ with the guaranteed result 1 % is obviously worse than the risky portfolio $\mathbf{x}''$ giving as possible returns 3 % and 5 %. Actually, $R_{\mathbf{x}''} \succ_{FSD} R_{\mathbf{x}'}$ and no rational investor would prefer $\mathbf{x}'$ to $\mathbf{x}''$. This weakness may be overcome by the use of safety measures instead of dispersion type risk measures.

The SSD consistency of the safety measures may be formalized as follows. We say that the safety measure $\mu(\mathbf{x}) - \varrho(\mathbf{x})$ is *SSD consistent* or that the corresponding risk measure $\varrho(\mathbf{x})$ is *SSD safety consistent* if

$$R_{\mathbf{x}'} \succeq_{SSD} R_{\mathbf{x}''} \quad \Rightarrow \quad \mu(\mathbf{x}') - \varrho(\mathbf{x}') \geq \mu(\mathbf{x}'') - \varrho(\mathbf{x}''). \tag{6.7}$$

The relation of SSD (safety) consistency is called *strong* if, in addition to (6.7), the following holds

$$R_{\mathbf{x}'} \succ_{SSD} R_{\mathbf{x}''} \quad \Rightarrow \quad \mu(\mathbf{x}') - \varrho(\mathbf{x}') > \mu(\mathbf{x}'') - \varrho(\mathbf{x}''). \tag{6.8}$$

Certainly, the SSD consistency condition (6.7) is fulfilled by the CVaR measure (2.19) as representing the normalized value of the second order quantile function $F_{\mathbf{x}}^{(-2)}(\beta)$. It turns out, however, that many other LP computable safety measures are SSD consistent due to the following theorem.

Theorem 6.1 *If the risk measure $\varrho(\mathbf{x})$ is SSD safety consistent (6.7), then except for portfolios with identical values of $\mu(\mathbf{x})$ and $\varrho(\mathbf{x})$, every efficient solution of the bicriteria problem*

$$\max\{[\mu(\mathbf{x}), \mu(\mathbf{x}) - \varrho(\mathbf{x})] : \quad \mathbf{x} \in Q\} \tag{6.9}$$

is an SSD efficient portfolio. In the case of strong SSD safety consistency (6.8), every portfolio $\mathbf{x} \in Q$ efficient to (6.9) is, unconditionally, SSD efficient.

Following Theorem 6.1, all the LP computable safety measures we presented in Chap. 2 are SSD consistent. This applies to the safety measures built as corresponding to the originally considered risk measures as well as to the measures originally considered in the safety form. We list all these measures referring also to the original portfolio optimization models. The following LP computable safety measures are SSD consistent:

- The mean downside underachievement (2.31) – the safety version (2.32) of the semi-MAD model,
- The worst realization (2.13) – the Minimax model (2.14),
- The CVaR measure (2.19) – the CVaR model (2.20),
- The mean worse return (2.33) – the safety version (2.35) of the GMD model,
- The weighted CVaR measure (2.41) – the WCVaR model (2.43),
- The safety measure corresponding to the mean penalized semideviation (2.44) – the safety version of the m-MAD model (2.45),

- The safety measure corresponding to the downside Gini's mean difference (2.47) – the safety version (2.48) of the Downside GMD model,
- Any convex combination of the above listed measures.

Moreover, in the case of the safety version of the GMD measure (2.33) or any convex combination of the above containing this measure, the consistency is strong.

When analyzing the SSD safety, consistency results from the perspective of the corresponding mean-risk bicriteria optimization. The efficient portfolios of the mean-safety problem form a subset of the entire ϱ/μ-efficient set spanning from the best expectation portfolio (BEP) to the minimum risk portfolio (MRP) as illustrated in Fig. 1.7. The maximum safety portfolio (MSP) distinguishes a part of the mean-risk efficient frontier, from BEP to MSP, which is also mean-safety efficient. In the case of a SSD safety consistent risk measure, this part of the efficient frontier represents portfolios which are SSD efficient. More precisely, if a point (ϱ, μ) located at the efficient frontier between BEP and MSP is generated by a unique portfolio, then this portfolio is SSD efficient. In the case of multiple portfolios generating the same point in the ϱ/μ image space, at least one of them is SSD efficient, but some of them may be SSD dominated. The strong SSD consistency guarantees that all possible alternative portfolios are SSD efficient.

6.4 Coherent Measures

The risk measures we consider are defined as (real valued) functions of distributions rather than random variables themselves. Nevertheless, their properties as functions of random variables can be analyzed. The class of coherent risk measures is defined by means of several axioms and is widely recognized. The axioms depict the most important issues in the comparison of risky prospects. Let us consider a linear space of random variables $\mathcal{L}$. A real valued performance function $C : \mathcal{L} \to \mathbb{R}$ is called a *coherent risk measure* on $\mathcal{L}$ if for any $R_{\mathbf{x}'}, R_{\mathbf{x}''} \in \mathcal{L}$ it is

(i) Monotonous: $R_{\mathbf{x}'} \geq R_{\mathbf{x}''}$ implies $C(R_{\mathbf{x}'}) \leq C(R_{\mathbf{x}''})$,
(ii) Subadditive: $C(R_{\mathbf{x}'} + R_{\mathbf{x}''}) \leq C(R_{\mathbf{x}'}) + C(R_{\mathbf{x}''})$,
(iii) Positively homogeneous: $C(hR_{\mathbf{x}'}) = hC(R_{\mathbf{x}'})$ for any positive real number h,
(iv) Translation equivariant: $C(R_{\mathbf{x}'} + a) = C(R_{\mathbf{x}'}) - a$, for any real number a,
(v) Risk relevant: $R_{\mathbf{x}'} \leq 0$ and $R_{\mathbf{x}'} \neq 0$ implies $C(R_{\mathbf{x}'}) > 0$,

where all inequalities on random variables are understood in terms of almost sure (a.s.). Note that if the performance function is positively homogeneous, then the axiom of subadditivity is equivalent to the standard convexity requirement:

(iia) $C(\alpha R_{\mathbf{x}'} + (1-\alpha)R_{\mathbf{x}''}) \leq \alpha C(R_{\mathbf{x}'}) + (1-\alpha)C(R_{\mathbf{x}''})$ for any $0 \leq \alpha \leq 1$.

The requirement of subadditivity (or convexity) is crucial to guarantee that diversification does not increase risk. The last axiom (risk relevance) is, however, sometimes ignored or understood in a different way.

Note that typical dispersion type risk measures can define the corresponding coherent risk measures, provided that they are SSD safety consistent and positively homogeneous. Indeed, typical dispersion type risk measures are convex, positively homogeneous and translation invariant, which implies that the composite objective $-\mu(\mathbf{x}) + \varrho(\mathbf{x})$, i.e. with sign changed with respect to the corresponding safety measure, does satisfy the axioms of translation equivariance, positive homogeneity, subadditivity. Moreover, the SSD safety consistency implies the monotonicity and justifies the coherence. Actually, the following assertion is valid.

Theorem 6.2 *Let $\varrho(x) \geq 0$ be a convex, positively homogeneous and shift independent (dispersion type) risk measure. If the measure satisfies additionally the SSD consistency*

$$R_{x'} \succeq_{SSD} R_{x''} \Rightarrow \mu(x') - \varrho(x') \geq \mu(x'') - \varrho(x''),$$

then the corresponding performance function $C(R_x) = \varrho(x) - \mu(x)$ fulfills the coherence axioms.

Coming back to the most common LP computable risk measures mentioned with shown SSD safety consistency, one may easily notice that all are convex, positively homogeneous and translation invariant. Hence, the negative versions of the corresponding safety measures fulfil the coherence axioms. Therefore, all the LP computable safety measures we presented in Chap. 2, if taken negative to be minimized, are coherent risk measures.

We list all these safety measures again referring also to the original portfolio optimization models. The negative versions of the following LP computable safety measures are coherent:

- The mean downside underachievement (2.31) – the safety version (2.32) of the semi-MAD model,
- The worst realization (2.13) – the Minimax model (2.14),
- The CVaR measure (2.19) – the CVaR model (2.20),
- The mean worse return (2.33) – the safety version (2.35) of the GMD model,
- The weighted CVaR measure (2.41) – the WCVaR model (2.43),
- The safety measure corresponding to the mean penalized semideviation (2.44) – the safety version of the m-MAD model (2.45),
- The safety measure corresponding to the downside Gini's mean difference (2.47) – the safety version (2.48) of the Downside GMD model,
- Any convex combination of the above listed measures.

Generally, the stochastic dominance relation, as based on the distributions, is more subtle than the coherency requirement formulated on random variables. In particular, the a.s. monotonicity used as the first coherency axiom is much stronger than the FSD or SSD monotonicity. On the other hand, the SSD consistency does not require several axioms of the coherency. Actually, the Omega ratio measure (2.52) is SSD consistent while it is not positively homogeneous and thereby not coherent.

Table 6.1 Sample returns

Portfolio	Scenario 1 (%)	Scenario 2 (%)
$\mathbf{x}^o$	1.5	1.5
$\mathbf{x}'$	3.5	4.5
$\mathbf{x}''$	5.0	4.0

A practical consequence of the lack of SSD consistency or the lack of coherence can be illustrated by three portfolios $\mathbf{x}^o$, $\mathbf{x}'$ and $\mathbf{x}''$ with rates of return (given in percentage) under two equally probable scenarios (Table 6.1). Note that the risk-free portfolio $\mathbf{x}^o$ with the guaranteed return 1.5 % is obviously worse than the risky portfolios: $\mathbf{x}'$ giving 3.5 % or 4.5 % and $\mathbf{x}''$ giving 5.0 % or 4.0 %. Certainly, in all models consistent with the preference axioms of either coherence or SSD, portfolio $\mathbf{x}^o$ is dominated by both $\mathbf{x}'$ and $\mathbf{x}''$. When a dispersion type risk measure $\varrho(\mathbf{x})$ is used, then all the portfolios may be efficient in the corresponding mean-risk model. Unfortunately, it also applies to portfolio $\mathbf{x}^o$, since for each such measure $\varrho(\mathbf{x}') > 0$ and $\varrho(\mathbf{x}'') > 0$ while $\varrho(\mathbf{x}^o) = 0$. This is a common flaw of all mean-risk models where risk is measured with some dispersion measure (Markowitz-type models). Further, let us notice that $R_{\mathbf{x}''} \succeq_{SSD} R_{\mathbf{x}'}$ although $R_{\mathbf{x}''} \not\succeq R_{\mathbf{x}'}$. Hence, the SSD consistency of a model guarantees that $R_{\mathbf{x}''}$ will be selected while the coherence allows that either $R_{\mathbf{x}''}$ or $R_{\mathbf{x}'}$ may be selected (it only guarantees that $R_{\mathbf{x}^o}$ will not be selected).

6.5 Notes and References

The expected utility concept is originated from Bernoulli work in 1738. Neumann and Morgenstern (1947) introduced axioms for a rational investor, showing formally existence of a utility function allowing the comparison of uncertain prospects according to their expected utility (the expected utility hypothesis). The so-called dual theory of choice under risk or rank-dependent utility was independently introduced by Quiggin (1982) and Yaari (1987). While in the expected utility models risk averse preferences are commonly accepted, the so-called prospect theory introduced by Kahneman and Tversky (1979) extends the approach to risk attitude dependent on the value of returns or losses.

The concept of stochastic dominance was inspired by earlier work in the theory of majorization Hardy et al. (1934). In economics, stochastic dominance was introduced in the 1960s as Quirk and Saposnik (1962) considered the FSD relation and demonstrated the connection to utility functions. The SSD representation of the risk averse preferences was brought to economics by Hadar and Russell (1969), Hanoch and Levy (1969), Rothschild and Stiglitz (1970), and Bawa (1975). Whitmore (1970) introduced the Third-order Stochastic Dominance (TSD) while an extension to higher order, even for fractional order, was presented by Fishburn (1976). The classical Lorenz curve is a popular tool to explain inequalities in income economics as representing a cumulative population versus income curve. It was redefined by Gastwirth (1971) in terms of the inverse of general distribution functions. The absolute (or generalized) Lorenz curve representing denormalized values of the inverse cdf was introduced by Shorrocks (1983). The quantile representation of the SD was initially introduced by Levy and Kroll (1978). Ogryczak and Ruszczyński (2002b) formally showed the duality (conjugency) relation between the second order quantile function and the second order cdf, thus justifying the quantile SSD characteristic for arbitrary distributions. Wide presentations of stochastic dominance orders may be found in books Levy (2006) and Müller and Stoyan (2002).

The SSD consistency of the Lower Partial Moments was shown by Fishburn (1977). Yitzhaki (1982) demonstrated the SSD safety consistency of the Gini's mean difference while the strong consistency was shown by Ogryczak and Ruszczyński (2002b). Ogryczak and Ruszczyński (1999) showed the SSD safety consistency of the semi-MAD and the standard downside semi-deviation. These results were further extended (Ogryczak and Ruszczyński 2001) on higher order downside semi-deviations from the mean and the higher order stochastic dominance, respectively, showing in particular the TSD safety consistency of the standard downside semi-deviation.

The notion of coherent risk measures was introduced in Artzner et al. (1999) and became widely recognized. Coherency of the CVaR measure was shown by Pflug (2000). Theorem 6.2 and thereby coherence of the negative of all the major LP computable safety measures was shown in Mansini et al. (2007).

Chapter 7
Computational Issues

7.1 Introduction

In the previous chapters, we have presented several different models for portfolio optimization. All the models are Linear Programming (LP) or Mixed Integer Linear Programming (MILP) models. In this chapter, we address related computational issues.

LP is the most used optimization tool in practice. It is used in a variety of applications, not only in finance. The success of LP is due to the availability of commercial software packages capable of solving instances of large size. When an LP problem is formulated, the software finds an optimal solution. When the formulation requires variables that are naturally integer or binary, the problem becomes a MILP model, and in many cases, the commercial software can solve to optimality only instances of small or medium size. Although it is out of the scope of this book to introduce LP and MILP techniques and related computational aspects, we briefly discuss the solution of the models through out-of-the shelf software packages and provide, whenever possible, hints to researchers and practitioners facing the solution of LP/MILP portfolio problems.

To tackle large size MILP problems, one has to resort to heuristics. A variety of heuristic and metaheuristic schemes is available in the literature. Given a specific problem, an ad hoc heuristic, inspired by any of those schemes, may be designed and implemented. A huge number of papers has been published containing heuristics purposely designed for the solution of specific problems. This approach may, however, become an obstacle to portfolio optimization, because of the expertise required and the effort, in terms of time and cost. Moreover, rather minor variations of a problem may imply the re-design of the heuristic, and thus additional time and cost. General purpose methods with respect to specific algorithms have the advantage that can be applied to large classes of MILP problems. Such a general purpose heuristic should be *efficient*, that is able to solve a large size MILP problem

© Springer International Publishing Switzerland 2015
R. Mansini et al., *Linear and Mixed Integer Programming for Portfolio Optimization*, EURO Advanced Tutorials on Operational Research,
DOI 10.1007/978-3-319-18482-1_7

within a reasonable amount of time. Moreover, the method should be *effective*, that is the solution obtained should be close to an optimal one. The goal of finding heuristics that are both efficient and effective, joined with the availability of software capable of efficiently solving small MILP problems, has led to the design of methods that are called *matheuristics*. Such methods combine heuristic schemes with the solution of small MILP subproblems and rely on available software for the solution of the latter. We present a matheuristic, called Kernel Search, that is general and flexible, and requires very limited implementation effort.

In this chapter, we also cover the computational issue related to the data needed by a model. The models presented must be fed with data, the most crucial being the rates of return of the assets under the different scenarios. Although several techniques can generate such data, we just present some of them focusing our attention on the one based on historical realizations as the simplest and most frequently used. There is another reason that makes historical data desirable in portfolio optimization models. Thus, the type of generation technique adopted affects the number of scenarios that have to be generated, and in turn the time required to solve the portfolio optimization model, where the number of constraints is proportional to the number of scenarios. For this reason, we dedicate a section of this chapter to the problem of solving large scale LP models whose size is caused by several thousands of scenarios. Although solving large LP models can be done efficiently, thanks to the advances in computing capability, this is true provided that the number of constraints remains limited. When the number of scenarios, and consequently the number of constraints and slack variables associated with them, grows to the order of several thousands, then solving an LP problem may take several minutes or even hours. We show how, in such cases, the computational efficiency can be achieved solving the dual problem.

The last computational issue we discuss deals with the comparison and testing of different models. Different risk/safety performance measures may be chosen. Different ways to include the transaction costs in the model may be considered. Different constraints may be added. The question is: Which model is the most appropriate to represent the problem? How can different models be compared? Are there any recognized methodologies? A researcher may have innovative ideas and propose a new model for portfolio optimization. The model may adopt a new performance measure, may overcome limitations of the models previously presented in the literature, may have better or stronger properties. A computational study may be unnecessary in the cases where the value of the new model is justified on a methodological or theoretical ground. However, such computational study may in other cases be necessary and, for the case where it is non necessary, may complement the analysis. In the last part of the chapter, we discuss how a model can be validated.

7.2 Solving Linear and Mixed Integer Linear Programming Problems

Algorithms to solve small and medium size LP problems are usually based on the simplex method. For large scale LP problems interior point methods (also referred to as barrier methods) are more commonly used.

Although most software packages available for the solution of LP models implement the same basic methods, the efficiency of a software package depends on its implementation and enhancements. Thus, different software packages have different performance, that is the computational time required to solve an instance of an LP model does depend on the software used. However, it is not simple to assess which software package is the best because the LP instances may be very different, for example in terms of sparsity of the coefficient matrix, and some may be solved more efficiently by a certain software package and others by a different one. Several benchmark instances have been made available over the years by researchers and companies on which different software packages have been tested and compared. It is common to see that no software package dominates all the others on the tested instances. However, from the computational experience of the scientific community and the tests reported in a large body of literature a few software packages emerge as the most efficient ones currently. A similar situation holds for the MILP models. The only, but major, difference is that while instances of LP models of real world size can be solved in a reasonable amount of time – on the order of seconds or minutes – instances of MILP models are computationally much harder. The maximum size of a MILP model that can be solved to optimality within a short amount of time, on the order of seconds or minutes, depends on the specific instance but there is consensus that in most cases it is on the order of hundreds of variables and constraints. Nowadays, the size of the instances of LP and MILP models that can be solved to proved optimality is much larger than the size that could be solved only a couple of decades ago. We expect further improvements in the future, due to the technological advances in hardware and in scientific research.

Several modeling languages are available that allow a model developer to code complicated optimization models in a clear, concise, and efficient way. The most standard modeling languages are: AIMMS, AMPL, GAMS, and MPL. Models developed in a modeling language can then be solved with any of the commercial optimizers. The most popular and well-known commercial software for the solution of LP, MILP and other classes of mathematical programming problems are currently CPLEX, Xpress and Gurobi. IBM ILOG CPLEX Optimization Studio (referred to simply as CPLEX) was named after the simplex method, implemented in the C programming language, although today CPLEX also supports other types of mathematical optimization and offers interfaces other than C. It was originally developed by Robert Bixby and offered commercially starting in 1988 by CPLEX Optimization Inc., which was acquired by ILOG in 1997. ILOG was subsequently acquired by IBM in January 2009 and CPLEX continues to be developed under IBM. FICO Xpress Optimization Suite software (referred to as Xpress) is the second most used platform for the solution of LP and MILP problems. Finally, the Gurobi Optimizer (referred to as Gurobi) was founded in 2008. Open source optimizers are also available, the most popular being GLPK, LP-SOLVE, Coin-OR.

Given that the computational time heavily depends on the model, the specific data and the solver, we may in general say that most LP models for portfolio optimization can be solved to optimality on instances of practical size.

The maximum size of instances of MILP models for portfolio optimization that can be solved to optimality strongly depends on the specific model and instance solved. Although from the theoretical point of view most models are NP-hard, computational experience has shown that some can be solved to optimality on instances with up to some hundreds of assets. One may wonder whether the number of assets has a stronger impact than the number of scenarios, or whether the cardinality constraints make the model harder to solve than the threshold constraints. Although simple indications that are guaranteed to hold in all cases cannot be provided, our computational experience suggests some general guidelines:

- In MILP models, an increasing number of assets makes the model harder to solve than an increasing number of scenarios; in fact, in general it is the number of binary or integer variables that determines the computational time required;
- Most models with fixed transaction costs may be solved to optimality within a reasonable amount of time (on the order of hours) on instances with up to several hundreds of variables;
- Models that include cardinality constraints become harder to solve when the maximum number of assets allowed becomes small; whereas with a maximum number of assets equal to 20 the models can be solved within a short time with up to hundreds of assets, with this limit set to 10 the model may not be solved to optimality within several hours;
- MILP models with CVaR measure are harder to solve than the corresponding models with MAD measure;
- MILP models with the CVaR as performance measure become harder when the value of β increases.

One might consider the design of ad hoc exact algorithms for the solution of a specific portfolio problem, especially if it turns out that instances of such problem cannot be solved efficiently through a commercial or open source optimizer. Such an algorithm should take advantage of the special structure of the problem and, for this reason, be more efficient than the general purpose MILP optimizers. The high quality of the best optimizers, however, makes this goal more and more difficult to achieve. Thus, when a MILP problem cannot be solved to optimality, it may become necessary to make use of a heuristic.

7.3 A General Heuristic: The Kernel Search

It is out of the scope of this book to discuss general heuristic schemes for MILP problems or ad hoc heuristics for the various MILP problems presented. In this section, we present, however, the Kernel Search, a general matheuristic that can be applied to any of these MILP problems and possible variants.

We consider a MILP problem with several sets of variables (for example, x, y and z) and assume, without loss of generality, that it is a minimization problem. For the sake of simplicity, we rename the variables in such a way that x is the set of all variables that are directly associated with assets, that is variables indexed on the assets.

We refer to the MILP problem including all the assets, and thus all the associated variables, as the *original problem*. We call *restricted problem* the problem restricted to the variables associated with a subset of the assets. The original problem can be seen as the problem restricted to the complete set of assets. We present a basic version of the Kernel Search, that we call *Basic Kernel Search (BKS)*.

In the BKS a sequence of restricted problems is solved. Let $N = \{1, \ldots, n\}$ be the complete set of assets. We use the notation MILP(K) to indicate the MILP problem restricted to the set of assets $K \subseteq N$. The scope of the BKS is to identify the assets that would be selected in an optimal solution of the original problem and to solve the MILP problem restricted to those assets. In fact, due to the diversification effect, in the large majority of portfolio optimization problems, independently of how large the number of the available assets is, only a relatively small number – often less than 100 – are selected. Thus, while the MILP problem solved on the complete set of assets may be computationally hard, the MILP problem restricted to one hundred assets can in most cases be solved in a short computational time.

A *kernel* is a set of assets that are likely to be selected in an optimal solution of the original problem. A *bucket* is a set of assets that are not contained in the kernel. The BKS first sorts the assets according to their likelihood to be selected in an optimal portfolio. The first n_I assets are chosen to be in the initial kernel. The remaining assets are partitioned into buckets. The MILP problem restricted to the initial kernel is solved. Then, the kernel is iteratively revised and the MILP problem restricted to the revised kernel plus the assets of one bucket is solved. In Algorithm 1 we provide a scheme of the BKS.

Algorithm 1 General scheme of the Basic Kernel Search

1. Identify the initial kernel and organize the remaining assets in a sorted list of buckets.
2. Solve the MILP problem restricted to the initial kernel.
3. Repeat until a stopping criterion is met:

   (a) Revise the kernel;
   (b) Solve the MILP problem restricted to the current kernel plus the next bucket in the list;
   (c) Remove the bucket from the list.

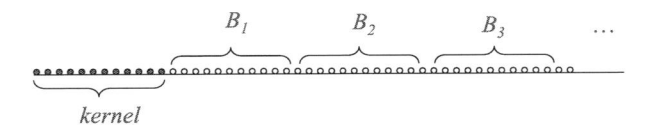

Fig. 7.1 Organization of the assets in the Basic Kernel Search

The steps (1) and (3a) of the BKS can be implemented in different ways. Now, we describe a possible implementation.

Step 1: We solve the continuous relaxation of MILP(N). If no feasible solution of the continuous relaxation exists, then no feasible solution of MILP(N) exists. Otherwise, the initial kernel is composed of the assets selected in an optimal solution of the continuous relaxation of MILP(N).

To create the sorted list of buckets, we have to first sort the assets that are not part of the initial kernel and then partition them into buckets. We consider two different situations. If the MILP problem is such that the continuous relaxation of MILP(N) provides reduced cost coefficients that are directly associated with the assets, we may use those coefficients to measure the likelihood of the assets to be in an optimal solution to the MILP problem. In this case, we sort the assets in non-decreasing order of the absolute value of the reduced cost. If the MILP problem is such that meaningful reduced cost coefficients are not available, then we may use any other criterion (such as Sortino ratio) to sort the assets. The idea is to find a criterion such that the assets which are more likely to belong to an optimal integer solution come first.

Then, let us choose a parameter l_B that is the number of assets in each bucket. The first l_B assets compose the first bucket, the subsequent l_B assets the second bucket and so on. The last bucket may contain less than l_B assets. Let N_B be the total number of buckets and B_i the set of assets in bucket i, $i = 1, \ldots, N_B$. In Fig. 7.1, we graphically represent how the assets are organized.

Step 3a: Each restricted MILP problem may find better solutions than the previous one and may identify further assets to insert into the kernel. If the latest restricted MILP problem solved has selected some assets from the current bucket, then the kernel is revised by including such promising assets.

The parameter l_B influences the size of the restricted problems solved. Its value should be the largest possible that allows us to solve in a reasonable amount of time the restricted MILP problems.

To speed up the solution of any restricted MILP problem, two additional constraints may be added:

- At least one asset from the current bucket must be selected; this constraint can be added because an optimal solution to the MILP problem restricted to the current kernel has been found at the previous iteration;
- The optimal value of the objective function must be better than the value of the best current feasible solution.

Indeed, with the introduction of these two constraints we are interested in those solutions that improve the previous best feasible solution value by using at least one new asset from the current bucket. Note also that with these two constraints the restricted MILP problem may turn out to be infeasible. If this is the case the procedure then proceeds with a new bucket.

It is crucial for the success of the BKS that each restricted MILP problem considers most of the assets that will be part of an optimal portfolio.

Several variants of the BKS may be thought of that may improve the basic version. For example, if the number of assets selected in the continuous relaxation of the MILP problem is small, the initial kernel may also contain assets that are not selected in the continuous relaxation. Also, as at the end of the BKS the kernel is different from the initial kernel and, most likely, closer to the set of assets of an optimal solution of MILP(N), one might consider that kernel as the initial kernel of a new run of the BKS. In the presented BKS the size of the kernel always increases. Usually, this does not create computational difficulties because, as already mentioned, thanks to the diversification effect, the number of assets in the kernel remains relatively small. In the cases where the number becomes large, one may also consider removing from the kernel the assets that are not selected by the restricted MILP problems.

Finally, it is obvious how the implementation effort required by the BKS is extremely small, as the BKS is heavily based on the availability of a software for the solution of MILP problems.

7.4 Issues on Data

In order to be solved, models need to be fed with data, and the quality of the data determines the quality of the output of any model. The most important data, needed in all the presented models, are the rates of return of the assets under the different scenarios. The process of generating such rates is called *scenario generation*. Historical data are the basis of most techniques for scenario generation. They are also used to test the performance of models. The data used to feed a model are called *in-sample*, whereas those used to assess the model performance are called *out-of-sample*. In general, different in-sample and out-of-sample data sets should be used. Four data sets corresponding to different market conditions, represented by a market index (FTSE 100), were introduced in Guastaroba et al. (2009b). Each data set temporal positioning is shown in Fig. 7.2. The first data set is characterized by an increasing market trend in the in-sample period as well as in the out-of-sample period (up-up trend), the second data set by an increasing trend in the in-sample period and by a decreasing one in the out-of-sample period (up-down trend), the third data set by a decreasing trend in the in-sample period and by an increasing one in the out-of-sample period (down-up trend) and, finally, the last set by a decreasing trend in both the in-sample and the out-of-sample periods (down-down trend). Each of these data sets consists of 2 years of in-sample weekly observations

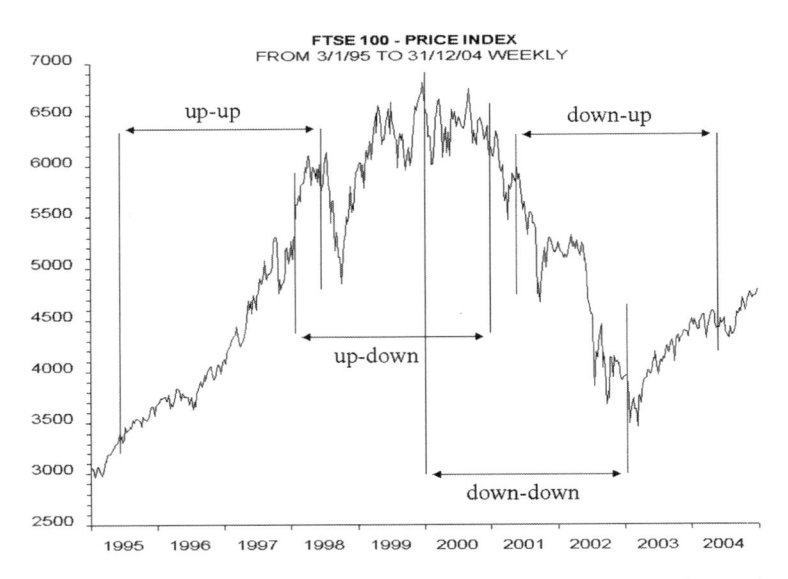

Fig. 7.2 Example of data sets combining different market trends (Source Guastaroba et al. (2009b))

(104 realizations) and 1 year of out-of-sample ones (52 realizations). This is a good way to generate in-sample and out-of-sample data: including and combining all possible market trends we guarantee a clear and thorough model validation.

The in-sample data are known at the time the investor builds the portfolio, whereas the out-of-sample data become available after the portfolio has been built. An investor builds his/her portfolio on the basis of the in-sample data, but is interested in the portfolio performance over the out-of-sample period. If the in-sample and out-of-sample data are consistent, in the sense that they show the same behavior (the up-up and down-down trends), we may expect a model to perform better than in the case the market behavior changes out-of-sample with respect to in-sample (the up-down and down-up trends). Tests performed on data sets that present the different trends allow a complete testing of a model. It is worth observing that, while in the past access to historical values of the asset returns was possible only through expensive databases, nowadays data can be easily found on the web at no cost (see, for example, www.finance.yahoo.com or www.google.com/finance).

We present here some of the techniques that can be adopted for scenario generation:

- *Historical data.* This method is one of the most frequently used for its simplicity. It is based upon the assumption that historical data represent possible future scenarios. A scenario corresponds to the realization of the rates of return of all assets as observed in a past period of time, e.g. a day or a week. Usually scenarios are treated as equally probable. Such approach does not require any assumption on the distribution function of the rates of return (the method is non-parametric).

7.4 Issues on Data

A potential drawback of this approach, and of all approaches making use of historical data, is that future scenarios may happen to be substantially different from those observed in the past. A relatively small number T of scenarios is sufficient to be significant, say on the order of hundreds. Thus, if the week is chosen as unit time period, in order to have 100 scenarios one has to go back 2 years in time to collect data. In a buy-and-hold strategy, taking the week as unit time period and considering 2 years of history represents a sensible way to generate scenarios.

- *Bootstrapping technique.* The method combines historical data with a bootstrapping technique. Bootstrapping is a re-sampling methodology based upon repeated sampling from the same set of data. As for the previous technique a scenario corresponds to the joint realization of the rates of return for all assets as observed in a given time period. The original historical set of scenarios is enlarged by sampling with replacement as many times as desired. The approach preserves the correlations between assets. The number of generated scenarios T is the sample size after re-sampling. The bootstrapping technique is frequently used when the size of the available sample is relatively small and one needs a larger number of scenarios.

- *Block bootstrapping technique.* Dealing with time series is more complicated than dealing with independent observations. The block bootstrapping technique is a variant of the previous one and allows us to retain original data correlations between periods by bootstrapping blocks of scenarios. A block is a set of consecutive scenarios. Given the original time series (historical data) and for a chosen block length, the method, instead of single scenarios, re-samples blocks of scenarios of the same length from the original set, thus retaining the correlations within the same block. The idea is to choose a block length large enough to guarantee that observations in different blocks are nearly independent. The number T of generated scenarios is the cardinality of the scenarios after the re-sampling.

- *Monte Carlo simulation techniques.* Monte Carlo simulation is an approach that consists in generating a large number of scenarios according to a specified distribution. The first critical issue when using this method as scenario generation technique in a portfolio optimization problem is the choice of the multivariate function which represents the distribution of the rates of return. The most frequently adopted distribution function is the multivariate standard normal distribution with zero mean and unit variance-covariance matrix, but other distribution functions may also be used. Since the multivariate normal distribution does not consider the "fat-tails" or "heavy-tails" effect which frequently characterizes rates of return, the multivariate t–Student distribution may be considered as a possible alternative.

- *Multivariate Generalized ARCH process technique.* Volatility cluster effect is a typical feature characterizing financial time series. Such effect is related to the observation that volatility for financial time series is usually not constant over time, but large rates of return tend to be followed by other large values, of both signs, while small rates of return are usually followed by small values. The simplest stochastic process taking into account such effect is the GARCH(q, p) (Generalized ARCH) stochastic process, where q and p are the number of the required past residuals and of the past conditional volatility values, respectively. For a given random variable, the residual between two consecutive time periods is computed as the difference between its expected value in the first period and its realization in the subsequent one. The original univariate GARCH(q, p) stochastic process can be extended to the multivariate case (M-GARCH(q, p)). Usually, the number of parameters to be estimated in a conventional M-GARCH(q, p) is too large and hence various approximations have been proposed instead.

Due to its computational simplicity, the Constant Conditional Correlation GARCH is widely used to estimate the multivariate GARCH stochastic process starting from univariate GARCH processes. We refer to this approach as M-GARCH(1, 1)(T), since we have set parameters p and q to 1, while T is the number of generated scenarios. The M-GARCH(1,1)(T) stochastic process can be used as a scenario generation technique through the construction of an event tree. Each node in the tree is associated with a joint outcome of the rates of return of all assets. An edge between two consecutive nodes means the child node was derived from the parent node according to the implemented stochastic process. Each parent node may generate more than one child node. A scenario is associated with each leaf of the tree. When generating the event tree a crucial issue is related to the number of nodes and stages to generate. One should find the right trade-off between a number of scenarios large enough to be significant and small enough to be computationally tractable.

The first three described scenario generation techniques are non-parametric, while the last two are parametric. The basic difference between a non-parametric and a parametric approach is that in the former case scenarios correspond to real data, while in the latter scenarios are based on assumptions on parameters and distributions. The non-parametric approaches have the main advantage to be simple to implement and easy to understand. On the contrary, a parametric approach depends on the chosen distribution. Moreover, there is no evidence that the distribution that best fits the data will be the one fitting in the future. On the other hand, non-parametric techniques are criticized to be strongly dependent upon a defined sample.

When choosing a technique for scenario generation one is concerned about how effective a technique is when embedded into portfolio optimization problems. Another important aspect concerns the computational burden implied by the different techniques, especially when compared to their effectiveness. We report here some results presented in Guastaroba et al. (2009b).

Table 7.1 shows the number of scenarios generated by the different implemented parametric and non-parametric techniques. The first line reports the name of the technique. In particular, *M-Norm(T)* and *M-tStud(T)* refer to the Monte Carlo simulation technique, under a multivariate normal and t-Student distribution,

7.4 Issues on Data

Table 7.1 Number of scenarios (sample size) generated with each technique (Source Guastaroba et al. (2009b))

Hist. data	Boot(T)	Block-Boot(T)	M-Norm(T)	M-tStud(T)	M-GARCH(1,1)(T)
104	1,000	1,000	1,000	1,000	1,024
	5,000	5,000	5,000	5,000	4,096
	10,000	10,000	10,000	10,000	8,192

Table 7.2 Average computational times (in minutes) to generate instances (Source Guastaroba et al. (2009b))

Sample size	Hist. data	Boot(T)	Block-Boot(T)	M-Norm(T)	M-tStud(T)	M-GARCH(1,1)(T)
104	0.17	–	–	–	–	–
1,000	–	0.42	0.50	5	20	60 (2^{10})
5,000	–	1.35	1.43	14	29	900 (2^{12})
10,000	–	2.50	2.67	25	40	3,480 (2^{13})

respectively, where T is the number of generated scenarios. For instance, column 2 refers to the bootstrapping technique indicated as Boot(T). The column values indicate that T was set to 1,000, 5,000 and 10,000, respectively, e.g. Boot(1,000) means that 1,000 scenarios were generated. For all the techniques but for the historical data and for M-GARCH(1,1)(T) the same number of scenarios was chosen. In the case of historical data sampling (Hist. Data), there are 104 historical scenarios. For the M-GARCH(1,1)(T) approach the number of scenarios has to be a power of 2 and, thus, T was set equal to 1,024, 4,096 and 8,192, respectively.

Table 7.2 shows, for each scenario generation technique, the average computational time required to build an instance as a function of the number of scenarios. The average times are expressed in minutes. Each row corresponds to a sample size and each row to a technique. For the M-GARCH(1,1)(T) model, the exact size of each sample is indicated in parentheses. Notice that, in the worst case, the non-parametric approaches require a computational time that is less than 3 min. Generating scenarios with both Monte Carlo simulation techniques is more time consuming (always more than 5 min). The most time consuming technique is the M-GARCH(1,1)(T). To create a binomial tree which consists of 10 stages one has to generate $\sum_{i=1}^{10} 2^i$ nodes. This required, even for the smallest instances, more than 1 h.

In Figs. 7.3 and 7.4, we show the performance (calculated in terms of cumulative rates of return) of a portfolio built at time 0 by using all described techniques with data sets corresponding to different trends (down-down and up-down). In fact, in each figure, the four methods that performed best are shown and their performance is compared to the market benchmark FTSE100. Considering the trade-off between the effort to generate the scenarios and the performance of the various techniques, the method based on historical data turns out to be a sensible choice. Although in some situations other methods perform better, it shows a stable and good average

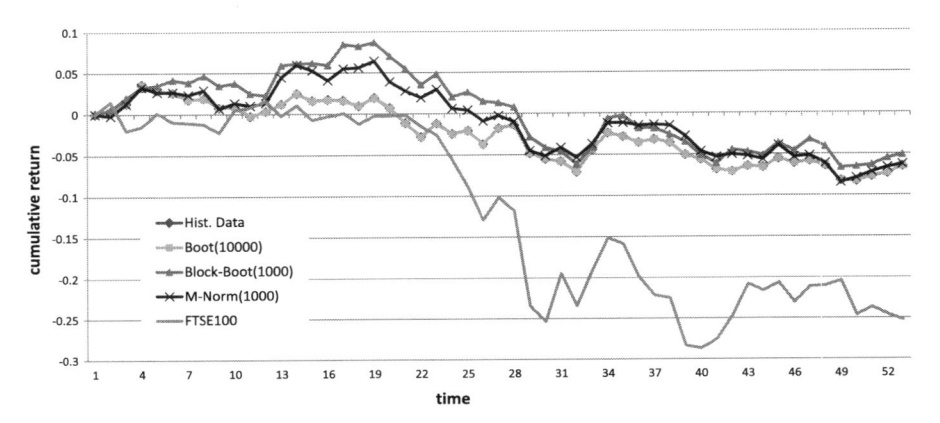

Fig. 7.3 Comparison of scenario generation techniques in down-down trend (Source Guastaroba et al. (2009b))

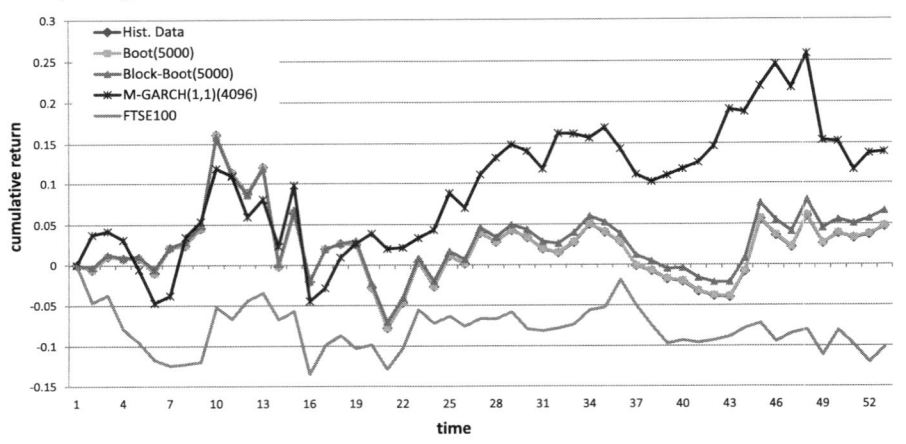

Fig. 7.4 Comparison of scenario generation techniques in up-down trend (Source Guastaroba et al. (2009b))

performance. This method is one of the most frequently used in the scientific literature.

7.5 Large Scale LP Models

Most of the LP models for portfolio optimization have dimensionality proportional to the number of scenarios T. For instance, the CVaR portfolio optimization model (2.20) can be formulated in its most compact form as:

$$\max \left(\eta - \frac{1}{\beta} \sum_{t=1}^{T} p_t d_t^- \right) \tag{7.1a}$$

7.5 Large Scale LP Models

$$d_t^- \geq \eta - \sum_{j=1}^{n} r_{jt}x_j \qquad t = 1, \ldots, T \qquad (7.1b)$$

$$\sum_{j=1}^{n} r_j x_j \geq \mu_0 \qquad (7.1c)$$

$$\sum_{j=1}^{n} x_j = 1 \qquad (7.1d)$$

$$d_t^- \geq 0 \qquad t = 1, \ldots, T \qquad (7.1e)$$

$$x_j \geq 0 \qquad j = 1, \ldots, n, \qquad (7.1f)$$

where η is an unbounded variable. The LP model (7.1) contains $T + n + 1$ variables and $T + 2$ constraints. Hence, the number of structural constraints (matrix rows) is proportional to the number of scenarios T, while the number of variables (matrix columns) is proportional to the total of the number of scenarios and the number of assets $T + n$. It does not cause any computational difficulty to have a few hundreds scenarios as in the case of historical data. However, with parametric scenario generation techniques one may have several thousands scenarios, thus leading to an LP model with a huge number of variables and constraints and thereby hardly solvable by means of general LP solvers, even when based on the interior point method. The computational efficiency can then be achieved by taking advantage of the LP dual of model (7.1) that takes the following form:

$$\min (v - \mu_0 u_0) \qquad (7.2a)$$

$$v - r_j u_0 - \sum_{t=1}^{T} r_{jt} u_t \geq 0 \qquad j = 1, \ldots, n \qquad (7.2b)$$

$$\sum_{t=1}^{T} u_t = 1 \qquad (7.2c)$$

$$0 \leq u_0, \ 0 \leq u_t \leq \frac{p_t}{\beta} \qquad t = 1, \ldots, T, \qquad (7.2d)$$

where v is an unbounded variable. The dual LP model contains $T + 1$ variables u_t, while the T constraints corresponding to variables d_t from (7.1) take the form of simple upper bounds on u_t (for $t = 1, \ldots, T$), thus not affecting the problem complexity. Actually, the number of constraints in (7.2) is independent from the number of scenarios. Exactly, there are $T + 1$ variables and $n + 1$ constraints. This guarantees an excellent computational efficiency even for a very large number of scenarios.

Similar reformulations can be applied to the classical LP portfolio optimization model based on the MAD measure as well as to more complex quantile risk measures. In the dual model the number of structural constraints is proportional to the number of assets n while only the number of variables is proportional to the number of scenarios T.

The standard LP model for the Gini's mean difference requires T^2 auxiliary constraints which makes it hard already for a medium number of scenarios, like a few hundreds scenarios given by historical data. Indeed, the standard GMD portfolio optimization model (2.12) formulated in the most compact form:

$$\max \ -\sum_{t=1}^{T} \sum_{t' \neq t} p_t p_{t'} d_{tt'} \tag{7.3a}$$

$$d_{tt'} \geq \sum_{j=1}^{n} r_{jt} x_j - \sum_{j=1}^{n} r_{jt'} x_j \qquad t, t' = 1, \ldots, T; \ t \neq t' \tag{7.3b}$$

$$\sum_{j=1}^{n} r_j x_j \geq \mu_0 \tag{7.3c}$$

$$\sum_{j=1}^{n} x_j = 1 \tag{7.3d}$$

$$d_{tt'} \geq 0 \qquad t, t' = 1, \ldots, T; \ t \neq t' \tag{7.3e}$$

$$x_j \geq 0 \qquad j = 1, \ldots, n, \tag{7.3f}$$

contains $T(T-1)$ nonnegative variables $d_{tt'}$ and $T(T-1)$ inequalities to define them. However, similarly to the CVaR models, variables $d_{tt'}$ are associated with the singleton coefficient columns. Hence, while solving the dual model instead of the original primal one, the corresponding dual constraints take the form of Simple Upper Bounds (SUB) which are handled implicitly outside the LP matrix. The dual of model (7.3) takes the following form:

$$\min \ (v - \mu_0 u_0) \tag{7.4a}$$

$$v - r_j u_0 - \sum_{t=1}^{T} \sum_{t' \neq t} (r_{jt} - r_{jt'}) u_{tt'} \geq 0 \qquad j = 1, \ldots, n \tag{7.4b}$$

$$0 \leq u_0, \ 0 \leq u_{tt'} \leq p_t p_{t'} \qquad t, t' = 1, \ldots, T; t \neq t', \tag{7.4c}$$

where the original portfolio variables x_j are dual prices associated with the inequalities (7.4b). The dual model contains $T(T-1)$ variables $u_{tt'}$ but the number of constraints (excluding the SUB structure) n is proportional to the number of assets. The above dual formulation can be further simplified by introducing variables $\bar{u}_{tt'} = u_{tt'} - u_{t't}$ for $t < t'$, which allows a reduction to $T(T-1)/2$ variables.

Thus, the models taking advantage of the LP duality allow one to limit the number of structural constraints making it proportional to the number of scenarios T and, therefore, increasing dramatically the computational performance for medium and large numbers of scenarios. Nevertheless, the case of a really large number of scenarios still may cause computational difficulties, due to a huge number of variables. The GMD risk measure may be then approximated with appropriately defined Weighted CVaR measures (2.40) defined as combinations of the CVaR measures for m tolerance levels. The corresponding LP portfolio optimization model (2.43) has the number of structural constraints equal to mT. When switching to the LP dual model the number of structural constraints is proportional to the total of the number of assets and the number of tolerance levels $n + m$. This guarantees a high computational efficiency of the dual model even for a very large number of scenarios.

It must be emphasized that the dual reformulation is applicable only to pure LP portfolio optimization models. It cannot be applied to the MILP models embedding various real features. This justifies the success of some measures as MAD and CVaR with respect to GMD when dealing with real features.

7.6 Testing and Comparison of Models

A model for portfolio optimization may be validated on the basis of two different lines of reasoning. The first is that the model has valuable/theoretical properties or captures relevant characteristics of a real problem. The second is that the model generates good portfolios, when implemented in a real-life framework. Often, even when a model is justified according to the first line of arguments, a computational validation is valuable.

Let us assume that the model we are interested in can be solved to optimality on the tested instances. The problem is how to show that the model generates good portfolios. One may build the efficient frontier, that is create for different values of the requested expected rate of return μ_0 the corresponding optimal portfolios. The optimal portfolios may be analyzed, for example, in terms of number and type of selected assets. This kind of analysis is called *in-sample*. In this section, to illustrate concepts, we make use of some of the results presented in Mansini et al. (2003a), where extensive computational results comparing LP models were reported. In Table 7.3, the average return of the MRP and MSP portfolios obtained using different models on two different data sets are shown. We recall that these are the portfolios that guarantee minimum risk and maximum safety, respectively, without any constraint on the portfolio expected rate of return. For each model the risk and the safety version were tested. The values were computed in-sample. One may say that, looking at the MSP columns, the MAD and Markowitz models are more aggressive than the others and that the CVaR model becomes more aggressive when the value of the parameter β increases.

Table 7.3 MRP and MSP average returns for two different data sets (in %) (Source Mansini et al. (2003a))

	Data set 1		Data set 2	
	MRP	MSP	MRP	MSP
Minimax	10.41	26.17	34.51	51.93
MAD	8.93	301.22	27.39	555.50
GMD	0.00	61.45	8.79	104.25
CVaR(0.1)	10.41	27.36	35.28	48.35
CVaR(0.5)	9.10	45.95	27.48	73.65
Markowitz	8.89	292.05	27.44	409.30

Table 7.4 Ranking of the first four assets (Source Mansini et al. (2003a))

	Asset 1		Asset 2		Asset 3		Asset 4	
Minimax	Bpcomin	(0.102)	Poligraf	(0.101)	Pininfrr	(0.096)	Cbarlett	(0.081)
MAD	Bayer	(0.128)	Bpcomin	(0.110)	Cbarlett	(0.097)	Bpintra	(0.071)
GMD	Bayer	(0.111)	Bpcomin	(0.110)	Cbarlett	(0.090)	Bpintra	(0.084)
CVaR(0.1)	Bpcomin	(0.105)	Cbarlett	(0.104)	Bplodi	(0.077)	Poligraf	(0.073)
CVaR(0.5)	Bayer	(0.124)	Cbarlett	(0.111)	Bpcomin	(0.095)	Premudar	(0.062)
Markowitz	Bayer	(0.141)	Crvaltel	(0.096)	Bpintra	(0.082)	Saesgetp	(0.076)

Table 7.5 Ranking of the four top assets (Source Mansini et al. (2003a))

	Bayer		Cbarlett		Bpcomin		Poligraf	
Minimax	No		0.089	(4)	0.102	(1)	0.101	(2)
MAD	0.128	(1)	0.0974	(3)	0.110	(2)	0.018	(18)
GMD	0.111	(1)	0.09	(3)	0.11	(2)	0.0314	(13)
CVaR(0.1)	0.0487	(10)	0.104	(2)	0.105	(1)	0.073	(4)
CVaR(0.5)	0.124	(1)	0.111	(2)	0.095	(3)	0.0197	(16)
Markowitz	0.141	(1)	0.0325	(13)	0.0578	(7)	0.0579	(6)

Another interesting analysis that allows a direct comparison of the various models concerns the portfolio composition in terms of chosen assets. In Table 7.4 the ranking of the four most selected assets is reported for a specific data set and for the case the required expected return was set to 17.5 % per year. In each cell the name of the asset and its share are shown.

The ranking of the assets selected by a specific model may vary for different levels of the required expected return, also for the same data set. Nevertheless, a core of the top ranked assets was observed to remain quite stable. It was also observed that the model CVaR(0.1) generates, on average, portfolios very similar to those selected by the Minimax model. Moreover, portfolios selected by the Markowitz model often contain a large number of assets with small shares.

In Table 7.5, the ranking of the four assets (namely, Bayer, Cbarlett, Bpcomin and Poligraf) with the largest share out of all the portfolios selected by the different models in the same period for a required return equal to 17.5 % per year is shown. The value of the share and the ranking position of the asset (if selected) in each

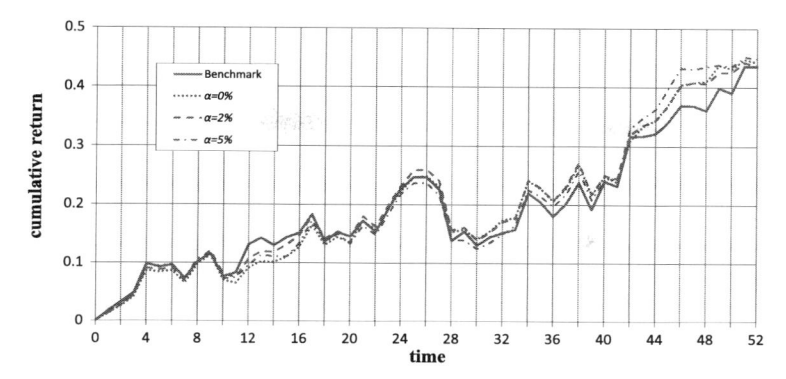

Fig. 7.5 Ex-post performance of an enhanced index tracking technique in up-up trend (Source Guastaroba et al. (2014))

portfolio are given. We can conclude that, with respect to the three most important assets in the portfolio ranking, the models tend to produce similar results. These three top assets cover, however, only about 30 % of the total investment.

The in-sample analysis is certainly interesting and valuable but what really matters is how a portfolio will perform in the future. This kind of analysis is called *ex-post*. The future returns of a portfolio, calculated on each unit time period or cumulated, may be compared with the returns of other portfolios or with market indices. In general, the ex-post performance of optimization models is strongly dependent on the data. Thus, to make it significant it should be performed on various data sets, representing different trends. As a general comment, we may say that more aggressive models tend to create portfolios that achieve higher average ex-post returns when the market increases, with returns that, however, vary substantially over time. This implies that, when the market decreases in the out-of-sample period, the more aggressive models tend to lose more.

As an example of ex-post analysis, in Fig. 7.5 (taken from Guastaroba et al. 2014) we show the ex-post performance of a model for enhanced index tracking over a period of 52 weeks. The parameter α associated with the different curves measures the level of over-performance of the model with respect to the benchmark. A higher value of α is associated with a higher requested level of over-performance.

7.7 Notes and References

A recent survey of LP solvers was presented by Fourer (2013). Several optimization resources and solvers are available at the NEOS server (http://www.neos-server.org/neos/) hosted by the University of Wisconsin in Madison.

The Kernel Search was introduced and applied to a combinatorial optimization problem by Angelelli et al. (2010). For applications to portfolio optimization and index tracking problems we refer to Angelelli et al. (2012) and Guastaroba and Speranza (2012), respectively.

The problem of efficiently solving large scale CVaR portfolio optimization models was approached in various ways including non-differential optimization techniques (Lim et al. 2010) or cutting planes procedures (Fabian et al. 2011). Ogryczak and Śliwiński (2011b) showed that the computational efficiency can be achieved with an alternative LP dual formulation. The efficiency of this approach was further confirmed by Espinoza and Moreno (2014). They showed that even more efficient primal-dual approaches exist but they need an implementation of specialized algorithms while the dual reformulation makes it possible to rely on standard solvers. Dual models for the MAD, MaxMin and Weighted CVaR portfolio optimization models were studied by Ogryczak and Śliwiński (2011a). Dual reformulation of the GMD and related portfolio optimization problems was introduced already in Krzemienowski and Ogryczak (2005).

For the analysis and comparison of different parametric and non-parametric techniques for scenario generation that can be seen as alternative techniques to historical data in LP models for portfolio optimization we refer to Kouwenberg and Zenios (2001) and Guastaroba et al. (2009b), and references therein. Several academic researchers and practitioners have used scenario generation techniques as tools for supporting financial decision making. In Pflug (2001) it was shown how a scenario tree may be constructed in an optimal manner on the basis of a simulation model of the underlying financial process by using a stochastic approximation technique. Finally, one of the most significant applications of scenario generation techniques based on historical data can be found in Cariño et al. (1994, 1998), where the authors developed the first commercial application making use of scenario generation methods for an asset allocation model applied to an insurance company.

References

Acerbi, C. 2002. Spectral measures of risk: A coherent representation of subjective risk aversion. *Journal of Banking & Finance* 26(7): 1505–1518.

Adcock, C., and N. Meade. 1994. A simple algorithm to incorporate transactions costs in quadratic optimisation. *European Journal of Operational Research* 79(1): 85–94.

Andersson, F., H. Mausser, D. Rosen, and S. Uryasev. 2001. Credit risk optimization with conditional value-at-risk criterion. *Mathematical Programming* 89(2): 273–291.

Angelelli, E., R. Mansini, and M.G. Speranza. 2008. A comparison of MAD and CVaR models with real features. *Journal of Banking & Finance* 32(7): 1188–1197.

Angelelli, E., R. Mansini, and M.G. Speranza. 2010. Kernel search: A general heuristic for the multi-dimensional knapsack problem. *Computers & Operations Research* 37(11): 2017–2026. Metaheuristics for Logistics and Vehicle Routing.

Angelelli, E., R. Mansini, and M.G. Speranza. 2012. Kernel search: A new heuristic framework for portfolio selection. *Computational Optimization and Applications* 51(1): 345–361.

Artzner, P., F. Delbaen, J.-M. Eber, and D. Heath. 1999. Coherent measures of risk. *Mathematical Finance* 9(3): 203–228.

Baumann, P., and N. Trautmann. 2013. Portfolio-optimization models for small investors. *Mathematical Methods of Operations Research* 77(3): 345–356.

Baumol, W.J. 1964. An expected gain-confidence limit criterion for portfolio selection. *Management Science* 10: 174–182.

Bawa, V.S. 1975. Optimal rules for ordering uncertain prospects. *Journal of Financial Economics* 2(1): 95–121.

Beasley, J.E., N. Meade, and T.-J. Chang. 2003. An evolutionary heuristic for the index tracking problem. *European Journal of Operational Research* 148(3): 621–643.

Bienstock, D. 1996. Computational study of a family of mixed-integer quadratic programming problems. *Mathematical Programming* 74(2): 121–140.

Canakgoz, N., and J. Beasley. 2009. Mixed-integer programming approaches for index tracking and enhanced indexation. *European Journal of Operational Research* 196(1): 384–399.

Cariño, D.R., T. Kent, D.H. Myers, C. Stacy, M. Sylvanus, A.L. Turner, K. Watanabe, and W.T. Ziemba. 1994. The Russell-Yasuda Kasai model: An asset/liability model for a Japanese insurance company using multistage stochastic programming. *Interfaces* 24(1): 29–49.

Cariño, D.R., D.H. Myers, and W.T. Ziemba. 1998. Concepts, technical issues, and uses of the Russell-Yasuda Kasai financial planning model. *Operations Research* 46(4): 450–462.

Chang, T.-J., N. Meade, J. Beasley, and Y. Sharaiha. 2000. Heuristics for cardinality constrained portfolio optimisation. *Computers & Operations Research* 27(13): 1271–1302.

Chekhlov, A., S. Uryasev, and M. Zabarankin. 2005. Drawdown measure in portfolio optimization. *International Journal of Theoretical and Applied Finance* 8(1): 13–58.

Chen, A.H., F.J. Fabozzi, and D. Huang. 2010. Models for portfolio revision with transaction costs in the mean–variance framework. In *Handbook of portfolio construction*, ed. John B. Guerard, 133–151. New York/London: Springer.

Chen, A.H., F.C. Jen, and S. Zionts. 1971. The optimal portfolio revision policy. *Journal of Business* 44(1): 51–61.

Chiodi, L., R. Mansini, and M.G. Speranza. 2003. Semi-absolute deviation rule for mutual funds portfolio selection. *Annals of Operations Research* 124(1–4): 245–265.

Elton, E., and M. Gruber. 1995. *Modern portfolio theory and investment analysis*, Portfolio management series. New York: Wiley.

Elton, E., M. Gruber, S. Brown, and W. Goetzmann. 2003. *Modern portfolio theory and investment analysis*. New York: Wiley.

Embrechts, P., C. Klüppelberg, and T. Mikosch. 1997. *Modelling extremal events: For insurance and finance*, Applications of mathematics. New York: Springer.

Espinoza, D., and E. Moreno. 2014. A primal-dual aggregation algorithm for minimizing conditional value-at-risk in linear programs. *Computational Optimization and Applications* 59(3): 617–638.

Fabian, C.I., G. Mitra, and D. Roman. 2011. Processing second-order stochastic dominance models using cutting-plane representations. *Mathematical Programming* 130(1): 33–57.

Feinstein, C.D., and M.N. Thapa. 1993. A reformulation of a mean–absolute deviation portfolio optimization model. *Management Science* 39: 1552–1553.

Fieldsend, J.E., J. Matatko, and M. Peng. 2004. Cardinality constrained portfolio optimisation. In *IDEAL*, Exeter, vol. 3177, ed. Z.R. Yang, R.M. Everson, and H. Yin. Lecture Notes in Computer Science, 788–793. Springer.

Fishburn, P.C. 1976. Continua of stochastic dominance relations for bounded probability distributions. *Journal of Mathematical Economics* 3(3): 295–311.

Fishburn, P.C. 1977. Mean-risk analysis with risk associated with below target returns. *American Economic Revue* 67: 116–126.

Fourer, R. 2013. Linear programming software survey. *OR/MS Today* 40(3): 40–53.

Gastwirth, J.L. 1971. A general definition of the lorenz curve. *Econometrica* 39(6): 1037–1039.

Guastaroba, G., R. Mansini, W. Ogryczak, and M. Speranza. 2014. Linear programming models based on Omega ratio for the enhanced index tracking problem. Tech. Rep. 2014–33, Institute of Control and Computation Engineering, Warsaw University of Technology.

Guastaroba, G., R. Mansini, and M.G. Speranza. 2009a. Models and simulations for portfolio rebalancing. *Computational Economics* 33(3): 237–262.

Guastaroba, G., R. Mansini, and M.G. Speranza. 2009b. On the effectiveness of scenario generation techniques in single-period portfolio optimization. *European Journal of Operational Research* 192(2): 500–511.

Guastaroba, G., and M.G. Speranza. 2012. Kernel search: An application to the index tracking problem. *European Journal of Operational Research* 217(1): 54–68.

Hadar, J., and W.R. Russell. 1969. Rules for ordering uncertain prospects. *American Economic Review* 59(1): 25–34.

Hanoch, G., and H. Levy. 1969. The efficiency analysis of choices involving risk. *Review of Economic Studies* 36(107): 335–346.

Hardy, G.H., J.E. Littlewood, and G. Pólya. 1934. *Inequalities*. London: Cambridge University Press.

Jobst, N., M. Horniman, C. Lucas, and G. Mitra. 2001. Computational aspects of alternative portfolio selection models in the presence of discrete asset choice constraints. *Quantitative Finance* 1(5): 489–501.

Jorion, P. 2006. *Value at risk: The new benchmark for managing financial risk*, 3rd edn. New York: Mcgraw-Hill.

Kahneman, D., and A. Tversky. 1979. Prospect theory: An analysis of decision under risk. *Econometrica* 47(2): 263–291.

References

Kellerer, H., R. Mansini, and M. Speranza. 2000. Selecting portfolios with fixed costs and minimum transaction lots. *Annals of Operations Research* 99(1–4): 287–304.

Konno, H., K. Akishino, and R. Yamamoto. 2005. Optimization of a long-short portfolio under nonconvex transaction cost. *Computational Optimization and Applications* 32(1–2): 115–132.

Konno, H., and A. Wijayanayake. 2001. Portfolio optimization problem under concave transaction costs and minimal transaction unit constraints. *Mathematical Programming* 89(2): 233–250.

Konno, H., and R. Yamamoto. 2005. Global optimization versus integer programming in portfolio optimization under nonconvex transaction costs. *Journal of Global Optimization* 32(2): 207–219.

Konno, H., and H. Yamazaki. 1991. Mean–absolute deviation portfolio optimization model and its application to tokyo stock market. *Management Science* 37: 519–531.

Koshizuka, T., H. Konno, and R. Yamamoto. 2009. Index-plus-alpha tracking subject to correlation constraint. *International Journal of Optimization: Theory, Methods and Applications* 1: 215–224.

Kouwenberg, R., and S. Zenios. 2001. Stochastic programming models for asset liability management. In *Handbook of asset and liability management*, ed. S. Zenios, W. Ziemba, 253–299. Amsterdam: North-Holland.

Krejić, N., M. Kumaresan, and A. Rožnjik. 2011. VaR optimal portfolio with transaction costs. *Applied Mathematics and Computation* 218(8): 4626–4637.

Krokhmal, P., J. Palmquist, and S. Uryasev. 2002. Portfolio optimization with conditional value-at-risk objective and constraints. *Journal of Risk* 4(2): 11–27.

Krzemienowski, A. 2009. Risk preference modeling with conditional average: Anăapplication to portfolio optimization. *Annals of Operations Research* 165(1): 67–95.

Krzemienowski, A., and W. Ogryczak. 2005. On extending the LP computable risk measures to account downside risk. *Computational Optimization and Applications* 32(1–2): 133–160.

Kumar, R., G. Mitra, and D. Roman. 2010. Long-short portfolio optimization in the presence of discrete asset choice constraints and two risk measures. *Journal of Risk* 13(2): 71–100.

Le Thi, H.A., M. Moeini, and T.P. Dinh. 2009. DC programming approach for portfolio optimization under step increasing transaction costs. *Optimization* 58(3): 267–289.

Lee, E.K., and J.E. Mitchell. 2000. Computational experience of an interior-point SQP algorithm in a parallel branch-and-bound framework. In *High performance optimization*, vol. 33, ed. H. Frenk, K. Roos, T. Terlaky, S. Zhang. Applied Optimization, 329–347. Boston: Springer US.

Levy, H. 2006. *Stochastic dominance: Investment decision making under uncertainty*, 2nd edn. New York: Springer.

Levy, H., and Y. Kroll. 1978. Ordering uncertain options with borrowing and lending. *Journal of Finance* 33(2): 553–574.

Li, D., X. Sun, and J. Wang. 2006. Optimal lot solution to cardinality constrained mean–variance formulation for portfolio selection. *Mathematical Finance* 16(1): 83–101.

Lim, C., H.D. Sherali, and S. Uryasev. 2010. Portfolio optimization by minimizing conditional value-at-risk via nondifferentiable optimization. *Computational Optimization and Applications* 46(3): 391–415.

Lintner, J. 1965. The valuation of risky assets and the selection of risky investments in stock portfolios and capital budget. *Review of Economics and Statistics* 47: 13–37.

Liu, S., and D. Stefek. 1995. A genetic algorithm for the asset paring problem in portfolio optimization. In *Proceedings of the first international symposium on operations research and its application (ISORA)*, Beijing, 441–450.

Lobo, M., M. Fazel, and S. Boyd. 2007. Portfolio optimization with linear and fixed transaction costs. *Annals of Operations Research* 152: 341–365.

Mansini, R., W. Ogryczak, and M.G. Speranza. 2003a. LP solvable models for portfolio optimization: A classification and computational comparison. *IMA Journal of Management Mathematics* 14: 187–220.

Mansini, R., W. Ogryczak, and M.G. Speranza. 2003b. On LP solvable models for portfolio optimization. *Informatica* 14: 37–62.

Mansini, R., W. Ogryczak, and M.G. Speranza. 2007. Conditional value at risk and related linear programming models for portfolio optimization. *Annals of Operations Research* 152: 227–256.

Mansini, R., W. Ogryczak, and M.G. Speranza. 2014. Twenty years of linear programming based portfolio optimization. *European Journal of Operational Research* 234(2): 518–535.

Mansini, R., W. Ogryczak, and M.G. Speranza. 2015. Portfolio optimization and transaction costs. In *Quantitative financial risk management: Theory and practice*, ed. C. Zopounidis and E. Galariotis, 212–241. Oxford: Wiley.

Mansini, R., and M.G. Speranza. 1999. Heuristic algorithms for the portfolio selection problem with minimum transaction lots. *European Journal of Operational Research* 114(2): 219–233.

Mansini, R., and M.G. Speranza. 2005. An exact approach for portfolio selection with transaction costs and rounds. *IIE Transactions* 37(10): 919–929.

Markowitz, H.M. 1952. Portfolio selection. *Journal of Finance* 7: 77–91.

Markowitz, H.M. 1959. *Portfolio selection: Efficient diversification of investments*. New York: Wiley.

Mausser, H., D. Saunders, and L. Seco. 2006. Optimising omega. *Risk Magazine* 19(11): 88–92.

Meade, N., and J.E. Beasley. 2011. Detection of momentum effects using an index out-performance strategy. *Quantitative Finance* 11(2): 313–326.

Michalowski, W., and W. Ogryczak. 2001. Extending the MAD portfolio optimization model to incorporate downside risk aversion. *Naval Research Logistics* 48(3): 185–200.

Mossin, J. 1966. Equilibrium in a capital asset market. *Econometrica* 34: 768–783.

Müller, A., and D. Stoyan. 2002. *Comparison methods for stochastic models and risks*. New York: Wiley.

Nawrocki, D.N. 1992. The characteristics of portfolios selected by n-degree lower partial moment. *International Review of Financial Analysis* 1(3): 195–209.

Neumann, J.V., and O. Morgenstern. 1947. *Theory of games and economic behavior*, 2nd edn. Princeton: Princeton University Press.

Ogryczak, W. 1999. Stochastic dominance relation and linear risk measures. In *Financial modelling – Proceedings of the 23rd meeting EURO WG financial modelling, 1998*, Cracow, ed. A.M. Skulimowski, 191–212. Progress & Business Publisher.

Ogryczak, W. 2000. Multiple criteria linear programming model for portfolio selection. *Annals of Operations Research* 97(1–4): 143–162.

Ogryczak, W., and A. Ruszczyński. 1999. From stochastic dominance to mean-risk models: Semideviations as risk measures. *European Journal of Operational Research* 116(1): 33–50.

Ogryczak, W., and A. Ruszczyński. 2001. On consistency of stochastic dominance and mean-semideviation models. *Mathematical Programming* 89(2): 217–232.

Ogryczak, W., and A. Ruszczyński. 2002a. Dual stochastic dominance and quantile risk measures. *International Transactions in Operational Research* 9(5): 661–680.

Ogryczak, W., and A. Ruszczyński. 2002b. Dual stochastic dominance and related mean-risk models. *SIAM Journal on Optimization* 13(1): 60–78.

Ogryczak, W., and T. Śliwiński. 2011a. On dual approaches to efficient optimization of LP computable risk measures for portfolio selection. *Asia-Pacific Journal of Operational Research* 28(1): 41–63.

Ogryczak, W., and T. Śliwiński. 2011b. On solving the dual for portfolio selection by optimizing conditional value at risk. *Computational Optimization and Applications* 50(3): 591–595.

Pflug, G.C. 2000. Some remarks on the value-at-risk and the conditional value-at-risk. In *Probabilistic constrained optimization: Methodology and applications*, ed. S. Uryasev, 272–281. Boston: Kluwer.

Pflug, G.C. 2001. Scenario tree generation for multiperiod financial optimization by optimal discretization. *Mathematical Programming* 89(2): 251–271.

Pogue, G.A. 1970. An extension of the Markowtiz portfolio selection model to include variable transaction costs, short sales, leverage policies and taxes. *Journal of Finance* 25(5): 1005–1027.

Quiggin, J. 1982. A theory of anticipated utility. *Journal of Economic Behavior & Organization* 3(4): 323–343.

Quirk, J.P., and R. Saposnik. 1962. The efficiency analysis of choices involving risk. *Review of Economic Studies* 29(2): 140–146.
Rockafellar, R., S. Uryasev, and M. Zabarankin. 2006. Generalized deviations in risk analysis. *Finance and Stochastics* 10(1): 51–74.
Rockafellar, R.T., and S. Uryasev. 2000. Optimization of conditional value-at-risk. *Journal of Risk* 2: 21–41.
Roman, D., K. Darby-Dowman, and G. Mitra. 2007. Mean-risk models using two risk measures: A multi-objective approach. *Quantitative Finance* 7(4): 443–458.
Rothschild, M., and J.E. Stiglitz. 1970. Increasing risk: I. A definition. *Journal of Economic Theory* 2(3): 225–243.
Roy, A. 1952. Safety-first and the holding of assets. *Econometrica* 20: 431–449.
Shadwick, W., and C. Keating. 2002. A universal performance measure. *Journal of Portfolio Measurement* 6(3 Spring): 59–84.
Sharpe, W.F. 1964. Capital asset prices: A theory of market equilibrium under conditions of risk. *Journal of Finance* 19: 425–442.
Sharpe, W.F. 1971a. A linear programming approximation for the general portfolio analysis problem. *Journal of Financial and Quantitative Analysis* 6: 1263–1275.
Sharpe, W.F. 1971b. Mean-absolute deviation characteristic lines for securities and portfolios. *Management Science* 18: B1–B13.
Shorrocks, A.F. 1983. Ranking income distributions. *Economica* 50(197): 3–17.
Smith, K.V. 1967. A transition model for portfolio revision. *Journal of Finance* 22(3): 425–439
Speranza, M.G. 1993. Linear programming models for portfolio optimization. *Finance* 14: 107–123.
Speranza, M.G. 1996. A heuristic algorithm for a portfolio optimization model applied to the Milan stock market. *Computers & Operations Research* 23(5): 433–441.
Stone, B.K. 1973. A linear programming formulation of the general portfolio selection problem. *Journal of Financial and Quantitative Analysis* 8: 621–636.
Tobin, J. 1958. Liquidity preference as behavior towards risk. *Review of Economic Studies* 25(2): 65–86.
Topaloglou, N., H. Vladimirou, and S.A. Zenios. 2002. CVaR models with selective hedging for international asset allocation. *Journal of Banking & Finance* 26(7): 1535–1561.
Valle, C., N. Meade, and J. Beasley. 2014a. Absolute return portfolios. *Omega* 45: 20–41.
Valle, C., N. Meade, and J. Beasley. 2014b. Market neutral portfolios. *Optimization Letters* 8: 1961–1984.
Whitmore, G.A. 1970. Third-degree stochastic dominance. *American Economic Review* 60(3): 457–459.
Woodside-Oriakhi, M., C. Lucas, and J. Beasley. 2013. Portfolio rebalancing with an investment horizon and transaction costs. *Omega* 41(2): 406–420.
Xidonas, P., G. Mavrotas, and J. Psarras. 2010. Portfolio construction on the Athens Stock Exchange: A multiobjective optimization approach. *Optimization* 59(8): 1211–1229.
Yaari, M.E. 1987. The dual theory of choice under risk. *Econometrica* 55(1): 95–115.
Yitzhaki, S. 1982. Stochastic dominance, mean variance, and Gini's mean difference. *American Economic Review* 72: 178–185.
Young, M.R. 1998. A minimax portfolio selection rule with linear programming solution. *Management Science* 44(5): 673–683.
Zenios, S., and P. Kang. 1993. Mean-absolute deviation portfolio optimization for mortgage-backed securities. *Annals of Operations Research* 45(1): 433–450.
Zhu, S., and M. Fukushima. 2009. Worst-case conditional value-at-risk with application to robust portfolio management. *Operations Research* 57(5): 1155–1168.

CPI Antony Rowe
Eastbourne, UK
May 15, 2020